Introduction to PHP for Scientists and Engineers

David R. Brooks

Introduction to PHP
for Scientists and Engineers

Beyond JavaScript

 Springer

David R. Brooks, PhD
Institute for Earth Sciences Research and Education
2686 Overhill Drive
Norristown, PA 19403
USA
brooksdr@drexel.edu

ISBN: 978-1-84800-236-4   e-ISBN: 978-1-84800-237-1
DOI 10.1007/978-1-84800-237-1

British Library Cataloguing in Publication Data
A catalogue record for this book is available from the British Library

Library of Congress Control Number: 2008926361

Printed on acid-free paper

9 8 7 6 5 4 3 2 1

Springer Science+Business Media
springer.com

Preface

The best way to become acquainted with a subject is to write a book about it.—Benjamin Disraeli

i. Background

The purpose of this book is provide an introduction to using a server-side programming language to solve some kinds of computing problems that cannot be solved with a client-side language such as JavaScript. The language is PHP (originally created in 1994 by Danish/Icelandic programmer Rasmus Lerdorf as "Personal Home Page Tools" for dealing with his own web site). The PHP language does not have a formal specification, as C does, for example. It is developed and maintained by a User Group of volunteers and is, essentially, defined by the most recently available free download. Although this might seem to be a shaky foundation on which to make a commitment to learning a programming language, PHP has a very large world-wide base of users and applications, which ensures its role into the foreseeable future.

This book should not be considered as a PHP reference source and it does not deal exhaustively even with those elements of the PHP language used in the book. (This should be considered a blessing by the casual programmer.) If you need more information, there is a huge amount of information online about PHP. Hopefully, this book will help you filter this information to focus on solving typical science and engineering problems. An excellent online source for information about PHP is http://www.php.net/manual/en/index.php, maintained by the PHP Documentation Group.[1]

This book is also definitely not intended as an introduction to programming. It is addressed not to professional programmers, actual or potential, but to a technical audience with occasional needs to solve computational problems. It assumes a working knowledge of programming concepts and HTML/JavaScript in particular, such as that provided by my book, *An Introduction to HTML and JavaScript for Scientists and Engineers* (Springer, 2007, ISBN-13: 978-1-84628-656-8,

---

[1] As with all URL references in this book, this URL was available at the time the book was written, but there can be no guarantee that it will exist in the future.

e-ISBN-13: 978-1-84628-657-5). Occasionally, this book uses examples drawn from the HTML/JavaScript book, but they are presented as stand-alone examples, so you do not need to own the HTML/JavaScript book to use this one.

Although PHP syntax is not that different from JavaScript, the file access syntax will be new to JavaScript programmers who have not previously programmed in a language such as C. C programmers will find there are some similarities between the file access syntax of PHP and C. However, the similarities are sometimes superficial and require that PHP be learned on its own terms, despite the temptation to conclude, "Oh, this works just like C (or JavaScript)."

As is often the case when I learn something new, I had a very specific goal when I started to learn about PHP: I needed to be able to create and access data files stored on a remote server. This is a capability that scientists and engineers always need, but which JavaScript simply cannot provide.

As with other texts I have written over the years, I followed the advice given in the quote at the beginning of this Preface. I wrote the book basically first for myself, and as a result this book will guide you through essentially the same steps I followed. As I began writing it, I had just finished publishing the HTML/JavaScript book mentioned above, so this book starts precisely where the previous one ended—with a knowledge of JavaScript sufficient for writing the online applications I needed, within the limitations of that client-side language.

Chapter 1 deals with the HTML/PHP interface that allows you to pass information from your local computer to a "remote" server, regardless of whether that server is really remote or exists on your own desktop. Chapter 2 deals with the basic elements of the PHP language, from the perspective of someone who is already familiar with programming concepts and JavaScript. If you have no idea what a "for loop" is, you will not be happy with this book!

Chapter 3 deals with arrays, which are a major topic because PHP's implementation of arrays is much richer and more complex than JavaScript's. Chapter 4 presents a summary of some elements of the PHP language used in the book. It is a *very* small subset of the language, but one that I have found meets my own needs. As you begin to apply this language to your own applications, you will no doubt want to make your own additions to my compilation of essentials. Chapter 5 provides a brief introduction to using PHP from a text-based command line interface. This bypasses the need to run PHP on a server and offers some basic facilities for providing user input to a PHP application from the keyboard.

There are many code examples in this book, most of them very short. They are "PHP applications" only in the most primitive sense. Their purpose is to illuminate a specific approach to a particular problem, such as reading a data file or producing properly formatted output. For anyone who, like me, does not make his or her living by programming and does not use a programming language every day, it is very easy to forget the details of how to achieve even simple tasks. I return to my own code examples again and again whenever I need to solve a new computing problem. So, I hope this book and its examples will save you as much time as they do me!

PHP is often used in conjunction with formal databases. However, that topic is not discussed at all in this book. Coming from a C background, I am used to more ad hoc and decidedly more primitive methods of creating and accessing my own data files. The specialized and constantly changing applications I need for my own work would only rarely benefit from a formal database structure.

Another omission that some programmers might find egregious is the lack of any mention of user-defined objects. Again, this is a decision based on my own needs. The additional capabilities that might be provided by creating objects are far outweighed by the practice and programming overhead required to use them correctly and efficiently. I believe this is equally true for this book's intended audience, as well.

A significant benefit of PHP relative to JavaScript is that, from a user's perspective, it appears to be a much more stable language. The language is supported by its own User Group, although it is not possible to predict the future of this support. Rather than being embedded in a browser, as JavaScript is, a PHP programming environment must be downloaded (at no cost) from essentially a single source, so it does not suffer from the many browser-dependent inconsistencies found in HTML/JavaScript.

Finally, a brief comment on the specialized nature of the main example addressed in the first two chapters: I have used it both because it was of interest to me when I first started to write this book and because I believe it represents a generic class of scientific and engineering computing problems. Because there are a lot of calculations, it provides many examples of how to use PHP's math functions; these, of course, are critical in science and engineering applications.

ii. Some typographic conventions used in this book

The code found in documents in this book is always copied directly from the editor used to create them (Visicomm Media's AceHTML freeware editor), as this is the best way to minimize errors and ensure that the code in the book actually works as intended. This editor displays some language elements in bold and/or italicized font and color-codes them, too. Of course, the color-coding is lost in this book, but the boldface and italicized fonts remain. When code examples in the text are not copied directly from AceHTML code, I have usually not bothered to reproduce that editor's font styles. In any case, those styles, including their presence or absence in the numerous code examples, have no significance relative to the PHP language.

Code and references to code elements in the text, such as function names, are always displayed in `Courier font`. In some code examples, user-supplied text is shown as *{Times Roman text enclosed in curly brackets}*. In some examples, including function parameter lists, variable names are given italicized "generic" names such as *$fileName* which the user can replace with some other application-specific name as needed.

Because of the format of this book, it was sometimes necessary to break lines of code in places where that wouldn't normally be necessary or desirable. I have tried to insert breaks in places where they won't create mischief with the code, but there may remain some cases where line breaks reproduced exactly as they appear in the text may produce a syntax error and prevent a script from executing.

The book contains a glossary with definitions of key terms, and the first appearance of each term appearing in the glossary is printed in **bold font**.

iii. Acknowledgments

I am yet again indebted to my wife Susan for her support and patient editing of this manuscript, especially considering how closely this chore followed my previous book on HTML and JavaScript. In the fall of 2007, I included PHP along with HTML and JavaScript in an introductory programming course I taught to graduate students in Drexel University's School of Biomedical Engineering, Science and Health Systems. Of course, it is always instructive to teach from a manuscript before submitting it, and I thank those students for their feedback!

Finally, I acknowledge relying on descriptions of PHP constructs and functions found online, in particular (but not exclusively) from the online PHP manual written and maintained by the PHP Documentation Group, found at us2.php.net/manual/en/. Especially when a succinct formal definition is available for a function or construct, from a definitive source, it is often difficult and pointless to try to be "original" about how such a description should be worded. However, I have tried to reword descriptions in a way that is consistent with the coverage provided in this book and with the needs of its intended audience, as I perceive it.

David R. Brooks

Institute for Earth Science Research and Education
January, 2008

Contents

1 Creating a Server-Side Environment for PHP   **1**

1.1 Getting Started   **1**
1.2 More Examples   **18**
    1.2.1 Solving the Quadratic Equation   18
    1.2.2 Preventing Multiple Submissions from a Form   20

2 Working with PHP   **23**

2.1 General Structure of PHP Scripts   **23**
2.2 Calculations with PHP   **24**
2.3 More about PHP File Input/Output   **40**
2.4 Another Example   **43**

3 PHP Arrays   **49**

3.1 Array Definition   **49**
3.2 Array Sorting   **54**
3.3 Stacks, Queues, and Line Crashers   **56**
3.4 More Examples   **59**
    3.4.1 The Quadratic Formula Revisited   59
    3.4.2 Reading `checkbox` Values   62
    3.4.3 Building a Histogram Array   66
    3.4.4 Shuffle a Card Deck   68
    3.4.5 Manage a Data File   69

4 Summary of Selected PHP Language Elements   **75**

4.1 Data Types and Operators   **75**
    4.2.1 Data Types   75
    4.2.1 Operators   76
4.2 Conditional Execution   **77**
    4.2.1 if... then... else... Conditional Execution   77
    4.2.2 Case-Controlled Conditional Execution   78
4.3 Loops   **79**
    4.3.1 Count-Controlled Loops   80
    4.3.2 Condition-Controlled Loops   82

4.4 Functions and Constructs   **83**
4.4.1 File Handling and I/O Functions   84
4.4.2 Math Constants and Functions   96
4.4.3 Array Functions and Constructs   99
4.4.4 Miscellaneous Functions and Constructs   102

5 Using PHP from a Command Line   *107*
5.1 The Command Line Environment   **107**
5.2 Is a Command Line Interface Useful?   **112**

Appendices   *115*
A.1 List of HTML and PHP Document Examples   **115**
A.2 ASCII Characters for Windows PCs   **117**

Exercises   *121*

Glossary   *133*

Index   *137*

1 Creating a Server-Side Environment for PHP

Chapter 1 compares the concepts of a server-side language such as PHP with the client-side environment provided by HTML and JavaScript. It shows how to set up an environment in which information from an HTML document can be sent to a PHP document located on a server which then can take actions based on that information.

1.1 Getting Started

JavaScript's primary limitation relative to languages such as C/C++ is that it is a **client-side**[1] language embedded in Web browsers. When an HTML/JavaScript document is accessed online or locally with a browser, only the contents of that document are available. JavaScript code cannot access data stored elsewhere on a **server**. This restriction is inherent in the language syntax and operating environment and applies regardless of whether the server is actually at a different location, a **remote server**, or whether external data merely exist elsewhere on a **local computer (server)** where the HTML/JavaScript document resides.

As a result, the only way to access information from a JavaScript application is to have all that information embedded within the application itself. Data can be formatted as a JavaScript array and contained in a file (often with a `.js` extension) that can be "included" when the HTML document is opened. This at least allows the data part of an application to be maintained separately from the processing part, but it does not overcome the inherent limitations of client-side processing.

As a result, because the ability to read and write data files is so important for science and engineering applications, it is necessary to use some other language in place of or in addition to JavaScript. One solution is to use PHP, a full-featured programming language similar in syntax to JavaScript and other languages derived from a C/C++ heritage. PHP is a

---

[1] Terns appearing in **bold font** the first time they are used are defined in the Glossary.

D.R. Brooks, *Introduction to PHP for Scientists and Engineers*,
doi: 10.1007/978-1-84800-237-1_1, © Springer-Verlag London Limited 2008

server-side language, which means that PHP documents reside on a remote or local server rather than being downloaded onto a client computer. Not only does PHP provide capabilities for accessing files stored on a remote or local server, it also allows an application to receive information from an HTML document and act on that information. For PHP to be used in this way, even on a local computer, PHP must be installed in an appropriately configured server environment, and not just installed as an application in some arbitrary folder.

When I first started to learn PHP, I hoped it would work just like a JavaScript function in the sense that I could send a PHP document some information, it would process that information, and finally it would "return" some results that could then be used within the HTML document. But PHP doesn't work this way! Passing information from an HTML document is strictly a one-way street, from client to server. The PHP document can display output on your browser screen, essentially by creating an HTML-formatted web page as output, but it doesn't return information that can be used within an HTML document or by a JavaScript script. For example, you cannot send information to a PHP application from the form fields in an HTML/JavaScript document, ask the PHP application to do some calculations, and then write the results back into form fields in the calling document for additional modifications through JavaScript. However, you *can* create and save data in files on the server, as will be shown later in this chapter.

As an example of how PHP works together with an HTML document, consider this problem:

> A user enters information about measurements taken with a sun-viewing instrument (called a sun photometer) that is used to measure total column water vapor in the atmosphere. The information provided by the user consists of the instrument's serial number, the location and time of the measurements, and voltage outputs from the instrument. The application must use this information to find the location of the sun at the time and place of the measurement and then calculate total column water vapor based on calibration constants stored for the instrument that collected the data.

The purpose of most of the calculations in this application will be first to determine the elevation (or zenith) angle of the sun based on the location and time of the measurements, to obtain what is called the relative air mass. Next, the calibration constants for the specified instrument must be retrieved. Finally, the precipitable water (PW) can be calculated. Apart from the actual calculations of solar position and PW, which are organizationally trivial, if computationally a little involved, this is a conceptually simple and typical data entry and processing problem that will provide a framework within which to learn about PHP.

Assume that the calibration constants (A, B, C, β, and τ) for several water vapor instruments are entered in a space-separated text file, WVdata.dat, and stored on a server:

```
SN A B C beta tau
WV2-113 0.762 0.468 0.20 0.65 0.10
WV2-114 0.814 0.468 0.20 0.650.10
...
WV2-157 0.911 0.468 0.20 0.65 0.10
...
```

A typical approach to programming problems is to separate a large problem into smaller problems. For this problem, the first step will be to write an HTML document that will pass an instrument serial number to a PHP application on a server. The second step is to write a server-side application that will receive this serial number and will then search through WVdata.dat and display the calibration constants for that instrument. These two documents must be linked through an interface that passes information from the first document to the second.

Here is an HTML document that will pass an instrument serial number to a server application. (Note: A complete list of all numbered documents in the book is found in Appendix A.1.)

Document 1.1. (getCalib.htm)

```html
<html>
<head>
<title>Get calibration constant</title>
<script language="javascript">
  document.write("This document last modified on "
    + document.lastModified+".")
</script>
</head>
<body>
<h2>Get calibration constants for water vapor
instrument</h2>
<p>
<form method="post" action="getCalib.php">
Enter serial number here: <input type="text" name="SN"
```

```
value="WV2-157" /><br />
<input type="submit"
  value="Click here to get calibration constants…" />
</body>
</html>
```

This document last modified on Thursday, February 01, 2007 11:05:02.

Get calibration constants for water vapor instrument

Enter serial number here: WV2-157

[Click here to get calibration constants…]

In this document, an instrument serial number is entered in typical fashion, as a value in a `type="text"` form field. A `type="submit"` form field button is used to send this value to a server-side application named `getCalib.php`. There is no reason why the PHP application needs to have the same name as the client-side HTML/JavaScript document, but using identical names (with different extensions) makes clear which HTML files are associated with which PHP files.

Note that JavaScript is used only incidentally in this document, to display the most recent document modification date and time. The transfer to a PHP document is done simply through the mechanism of an HTML form and does not require any other action such as a JavaScript script.

Information is sent to a PHP application simply by setting the `method` and `action` attributes in the `form` tag:

```
<form method="post" action="getCalib.php">
```

The value of the `action` attribute gives the location of the PHP document. The fact that, in this example, the value is simply the PHP file name implies that both the calling HTML document and the receiving PHP document reside in the same folder on a local computer—in this case, the folder designated as the local host on which PHP applications run. As will be demonstrated below, this action automatically transmits the name and value of every form field defined in the calling document. In this case, we are interested in only one value, the instrument serial number.

Now write the PHP application to receive and process this information. It is more complicated than Document 1.1 and ventures into territory that will be unfamiliar to JavaScript programmers.

Document 1.2. (`getCalib.php`)

```php
<?php
// Extract instrument ID from POST data...
  $SN=$_POST["SN"];
  $len=strlen($SN);
// Open WV instrument calibration constant file...
  $inFile = "WVdata.dat";
  $in = fopen($inFile, "r") or
     exit("Can't open file");
// Read one header line...
  $line=fgets($in);
// Search rest of file for SN match...
  $found=0;
  while ((!feof($in)) && ($found == 0)) {
// Could do it like this...
//      $line=fgets($in);
//      $values=sscanf($line,"%s %f %f %f %f %f");
    $values=fscanf($in,"%s %f %f %f %f %f");
    list($SN_dat,$A,$B,$C,$beta,$tau)=$values;
    if (strncasecmp($SN_dat,$SN,$len)==0) $found=1;
  }
  fclose($in);
  if ($found == 0)
    echo "Couldn't find this instrument.";
  else {
// Build table of outputs...
    echo "<p><table border='2'><tr><th>Quantity</th>
      <th>Value</th></tr>"."</td></tr>";
    echo "<tr><td>Instrument ID</td>
          <td>$SN</td></tr>";
    echo "<tr bgcolor='silver'>
      <td colspan='2'>Calibration Constants</td>
      </tr>";
    echo "<tr><td>A</td><td>$A</td></tr>";
    echo "<tr><td>B</td><td>$B</td></tr>";
    echo "<tr><td>C</td><td>$C</td></tr>";
    echo "<tr><td>&tau;</td><td>$tau</td></tr>";
    echo "<tr><td>&beta;</td><td>$beta</td></tr>";
    echo "</table>";
  }
?>
```

What is required to create and use such a document? The following steps proceed "from scratch," based on the assumption that you understand how to use HTML and JavaScript but have never used a server-side programming language.

Quantity	Value
Instrument ID	WV2-157
Calibration Constants	
A	0.911
B	0.468
C	0.2
τ	0.1
β	0.65

Step 1. Setting up a PHP environment.

There is a very significant difference between HTML/JavaScript documents and PHP applications. JavaScript is, essentially, universally and automatically available through any modern browser, so you shouldn't have to do anything special to process JavaScript code (unless, for some reason, your browser's JavaScript interpreter has been disabled). In contrast, the ability to use an HTML document as a source of input for a PHP application exists only on servers where a **PHP interpreter** has been installed. (PHP is an interpreted, as opposed to a compiled, language in the sense that stand-alone executable binary files are not generated. It is possible to produce such files, but this isn't normally necessary.) The server must be configured specifically to allow PHP scripts to be processed and all PHP applications must be saved on the server in an appropriate location. Because of the potential for carelessly written or malicious code to wreak havoc on any computer that allows remote access to its contents (in this case, through an HTML document that passes information to a PHP application), appropriate safeguards must be established to limit the ability to read and write data to specific locations on the server.

Even if the client browser and server reside on the same physical computer, a server and PHP interpreter must still be installed and configured, and precautions should still be taken to protect the computer's contents. On Windows computers, Microsoft's IIS (Internet Information Services) application[2] is a reasonable choice. This software should be included on the original operating system installation CD, but it may or may not have been installed at the time the OS was originally installed.

The steps involved in setting up a **PHP environment** for developing PHP applications on a local computer include:

---

[2] At the time this book was written, IIS was the name of this application under Windows XP. It may be different in subsequent operating systems. Other operating systems will use different servers.

1. Install a server on a local computer.

2. Download and install a PHP interpreter. At the time this book was written, this was a free download from http://www.php.net/downloads.php.

3. On a local computer, configure the server to recognize PHP documents and to locate the PHP interpreter.

Fortunately, there are many online sources of help with installing PHP on your own computer. For using PHP on a remote server, you don't have to carry out the above steps yourself, but you still need to know how to access the server. The details vary from system to system, and you may need to get help from your system administrator.

Step 2. Creating, editing, and executing PHP documents.

Just like HTML documents, **PHP documents** are text files that can be created with any text editor. For this book, I have used the same AceHTML freeware editor[3] that I use to create HTML/JavaScript documents on my Windows computers. AceHTML provides convenient editing and color-coded syntax formatting capabilities for creating PHP documents just as it does for creating HTML/JavaScript documents.

At first, it may not be obvious that you cannot execute PHP scripts in an editor's browser window, as you can JavaScript scripts in HTML documents. However, this is clearly the case. You can create and edit PHP scripts with an editor, but you must then execute them on a server, even if that server resides on your own computer. For example, on a Windows XP computer, with PHP installed on an IIS server, PHP documents will probably be saved in the `C:/Inetpub/wwwroot` folder and executed by entering

`localhost/{PHP document name}`

as the URL in a browser window. These folders are automatically created when the IIS software is installed. For convenience, you can also store the corresponding HTML/JavaScript documents in the same folder (recall Document 1.1).

So, to execute a PHP document, create it in a code or text editor, save it in `wwwroot`, then switch to a browser to execute it at `localhost`

---

[3] At the time this book was written, this editor was available as a free download from http://software.visicommedia.com/en/.

(in the case of a local Windows server). Whenever you make changes, save them and refresh the PHP document in your browser window.

When you create applications, whether in JavaScript, PHP, or some other language, it is important to develop a consistent approach that minimizes the time spent correcting the errors you will inevitably make. It is rarely a good idea to create an entire application all at once. A much better plan is to proceed in a step-by-step fashion, adding small sections of code and testing each addition to make sure the results are what you expect. When you pass information from an HTML document to a PHP application, it is *always* a good idea to display the values passed to the PHP application before writing more PHP code.

The error messages you will receive when you make mistakes in your code will almost never be very helpful, although experience and practice will improve your ability to interpret these messages. They may tell you where an error has been encountered but rarely what the error actually is ("You forgot to put a semicolon at the end of line 17."). PHP interpreters seem to be a little more helpful than JavaScript interpreters when it comes to describing errors, but neither of them will tell you what you really need to know—exactly what you did wrong and how to fix it. And of course, no syntax checker will protect you against the worst error of all—code that works perfectly well but is logically flawed and gives the wrong answers!

To test your PHP/server environment, start with this minimal PHP document. Name it `helloWorld.php` and save it in `wwwroot` (or the equivalent location on your system).

Document 1.3. (`helloWorld.php`)

```php
<?php
  echo "Hello, world!";
?>
```

Open a browser and type `localhost/helloWorld.php` (or whatever is the appropriate URL for your system). You should see the message, "Hello, world!" displayed in your browser window.

Another difference between HTML files and PHP files is that your computer is probably configured to automatically associate an `.htm` or `.html` extension with your browser. So, if you double-click on a file with such an extension, it will open in your browser. If you double-click on a file with a `.php` extension, its fate is less certain. For example, on a Windows computer, such files may opened as text files in the Notepad utility. In order to "execute" this file, you must enter its URL in a browser, as noted above.

You can also save and execute this file:

Document 1.4. (`PHPInfo.php`)

```php
<?php
  echo phpinfo();
?>
```

This file will display a great deal of information about how PHP is configured on your server. (If you view the source code for this document, you can also learn a lot about formatting output from PHP.)

The `echo` **language construct** in Document 1.3 displays the specified string literal and in Document 1.4 displays the string returned from the `phpinfo()` function.

The first thing to notice about these documents is that PHP files do not need to be embedded within an HTML document template, with its basic tags; they can also serve as stand-alone applications. Although you will often see PHP documents that place PHP scripts inside the `body` tag of an HTML document, these two examples demonstrate that PHP can work on its own without any HTML "shell," basically with the assumption that any HTML syntax appearing in formatted output will be interpreted appropriately when that output is echoed to your browser.

PHP code—a **PHP script**—is enclosed inside a **PHP tag**:

```php
<?php
  ...
?>
```

There are other ways of implementing PHP scripts, including

```
<script language="PHP">
  ...
</script>
```

but the `<?php ... ?>` tag found in the examples in this book is widely used for stand-alone PHP applications.

If you get error messages, or if nothing happens when you try to execute the documents in this chapter, then something is wrong with your server/PHP installation. It is hopeless to try to offer system-specific advice for resolving this kind of problem, but the most likely source of trouble at this level, assuming that you have installed both a server and

PHP, is that some server configuration options have been overlooked or have been given inappropriate settings, or that you haven't stored your PHP document in an appropriate folder on a local computer.

Step 3. Passing information from an HTML/JavaScript document to a PHP application.

The next thing you need to know to use PHP for anything other than displaying text output in a browser window is how to pass information to a PHP document. By design (since this is the essential reason for the existence of PHP), this is very easy to do, using the `action` and `method` attributes of the `form` element.

```
<form action="{URL of PHP document}" method="post">
...
<input type="submit" value="{Put submit button text here.}" />
</form>
```

The example from Document 1.1,

```
<form method="post" action="getCalib.php">
```

passes information to `getCalib.php`.

This book will always use `method="post"`, although it is also possible to use `method="get"` in some circumstances. A PHP document is identified by its URL and not by a directory/folder reference on the server. For the examples in this book, the assumption is that the HTML document and the PHP document reside in the same folder/directory on the same local server. For local use on a Windows computer running the IIS server, this location could be `C:/Inetpub/wwwroot` with a URL of `localhost`. This co-location of files is done just for convenience in a local server environment. When you use PHP on a remote server, you will store the HTML interface document on your local computer or download it from a server, and the URL for the PHP document will be different, of course.

What makes the process of calling a PHP document from an HTML document so painless is the fact that the contents of *all* form `input` fields in the calling document are *automatically* available to the target PHP document, without any additional programming effort on your part! There are some nuances and useful modifications to this very simple procedure, including ways to ensure that the "submit" action is carried out only once per visit to the calling document, but that discussion isn't essential for now.

On the **server side**, getCalib.php receives the information from getCalib.htm in an array named $_POST whose elements are accessed by using the names associated with the form fields in the calling document. In Document 1.2, the statements

```
$SN=$_POST["SN"];
$len=strlen($SN);
```

assign the value of the ID form field (its default value is WV2-157) to the PHP variable $SN and save the length of $SN in variable $len. In PHP scripts, all variable identifiers, whether user- or PHP-defined, begin with a $ character.

Step 4. Reading and interpreting information stored in server-side text files.

The next step in solving the problem addressed in Document 1.2 is to compare the value of $SN against the instrument serial numbers stored in WVdata.dat. Open the file and assign a name—a **file handle** in programming terminology:

```
$inFile = "WVdata.dat";
$in = fopen($inFile, "r") or exit("Can't open file.");
```

The file handle $in (it can be any name you like) provides a link between the physical file stored on the server and the "logical" name by which that file will be known in a PHP script. When it is first opened, the value of the file handle is the location in memory of the first byte of the physical file.

Files are opened with the fopen() function. The parameters are the file name and a string that specifies the operations that are allowed to be performed on the file. It is not necessary to assign a separate variable name to the physical file. The single statement

```
$in = fopen("WVdata.dat", "r") or
  exit("Can't open file.");
```

will work the same as the previous two statements, but the single "hard-coded" file name doesn't allow you to pass the name of the physical file from an HTML document to a variable in the PHP document. Of course, for some applications, you might *want* to use a hard-coded file name that can't be changed by a user calling the PHP document.

A value of "r" (or 'r') identifies the file as a **read-only text file**. This means that the PHP document can extract information from WVdata.dat but cannot change its contents in any way. The exit() function displays an appropriate message if the requested file doesn't exist, and exits the application.

There are several ways to read text files and extract information from them. The basic requirement is that the programmer must know *exactly* how data in the file are stored. The first line in WVdata.dat is a **header line**:

```
SN A B C beta tau
```

The header line is followed by calibration values for instruments. From the PHP script's point of view, the number of instrument calibrations stored in this file is unknown—additions to or deletions from the file can be made offline at any time. So, the script must first read past the header line (which is assumed always to be present in this application) and then search through an unknown number of instrument calibration data lines to find the specified instrument:

```
// Read one header line...
$line=fgets($in);
// Search rest of file for SN match...
$found=0;
while ((!feof($in)) && ($found == 0)) {
// Could do it like this...
// $line=fgets($in);
// $values=sscanf($line,"%s %f %f %f %f %f");
  $values=fscanf($in, "%s %f %f %f %f %f");
  list($SN_dat,$A,$B,$C,$beta,$tau)=$values;
  if (strncasecmp($SN_dat,$SN,$len)==0) $found=1;
}
fclose($in);
```

The syntax is similar to JavaScript syntax for conditional loops, even if the file-handling functions are unfamiliar because they have no equivalent in JavaScript. The variable $found is assigned an initial value of 0 before starting the loop. You could also assume that $found is a boolean value and initialize it with a value of false, to be changed later to true rather than 1, along with changing ($found == 0) to (!$found) in the while(...) statement.

Inside the loop, you can continue to read the file one line at a time. The `feof()` function terminates the conditional loop when an end-of-file mark is encountered, and the test on `$found` terminates the loop when the specified instrument serial number is found. Since every line is formatted the same way, the `fscanf()` function is a simple choice for extracting data from the file. The shaded line in the above code stores six values in the user-named array `$values`.

An alternative approach shown in the comment lines is to use the `fgets()` function again to read a line of text into a user-specified string variable called `$line`. The `sscanf()` is used to extract the values. For this application there is no reason to replace one statement for extracting values with two statements that accomplish exactly the same goal.

The **format string** in the `fscanf()` function tells PHP to look for a text string followed by five real numbers. (The "f" stands for **floating-point** number.) Each value in the text file must be separated by one or more spaces. This code will not work as written if other printable characters, such as commas, are present. However, extra spaces and even tabs are OK—tabs are treated as "white space" separating the values and are ignored by the **format** specification. Except for white space, the contents of the data file must *exactly* match what the format specifier string tells your code to expect.

The `list()` construct (which looks like a function because of the parentheses, but in programming terms is a language construct) associates the elements of `$values` with the values read from one line of data in the calibration file. The names of these values can be whatever you like. In this case, the choice that makes the most sense is to use the descriptive names that appear in the header line of the data file.

The `strncasecmp()` function performs a case-*in*sensitive comparison of the instrument serial number passed from `getCalib.htm` against serial numbers in the `WVdata.dat` file. (That is, a value of either `WV2-157` or `wv2-157` passed from `getCalib.htm` will be treated as a match with the `WV2-157` value in `WVdata.dat`. If a serial number match is found, `strncasecmp()` returns a value of 0 and the `$found` value is changed to 1. The `feof()` function looks for an "end-of-file" mark in the file and terminates the loop if it gets to the end of the file without finding a match with the user-specified serial number. When the loop terminates, the `fclose()` function closes the data file.

If a calibration for the specified instrument doesn't exist, then the script should print an appropriate message. Otherwise, it should display the calibration values for that instrument. Use an `if... else...` statement, with syntax similar to JavaScript:

```
if ($found == 0) echo "Couldn't find this
instrument.";
else {
// Build table of outputs...
  echo "<p><table
border='2'><tr><th>Quantity</th><th>Value</th></tr>";
  echo "</td></tr>";
  echo "<tr><td>Instrument ID</td><td>$SN</td></tr>";
  echo "<tr bgcolor='silver'>
     <td colspan='2'>Calibration Constants</td></tr>";
  echo "<tr><td>A</td><td>$A</td></tr>";
  echo "<tr><td>B</td><td>$B</td></tr>";
  echo "<tr><td>C</td><td>$C</td></tr>";
  echo "<tr><td>&tau;</td><td>$tau</td></tr>";
  echo "<tr><td>&beta;</td><td>$beta</td></tr>";
  echo "</table>";
}
```

This code shows how to use echo to build output strings that include HTML tags. Tags are used here to build a table, just as you would do in an HTML document. The multiple echo statements in this example could be reduced by using PHP's string concatenation operator, a period. The statements building the output table could also be written like this:

```
echo "<p><table border='2'>
. <tr><th>Quantity</th><th>Value</th></tr>"
. "</td></tr>"
. "<tr><td>Instrument ID</td><td>$SN</td></tr>"
. "<tr bgcolor='silver'><td colspan='2'>"
. "Calibration Constants</td></tr>"
. "<tr><td>A</td><td>$A</td></tr>"
. "<tr><td>B</td><td>$B</td></tr>"
. "<tr><td>C</td><td>$C</td></tr>"
. "<tr><td>&tau;</td><td>$tau</td></tr>"
. "<tr><td>&beta;</td><td>$beta</td></tr>"
. "</table>";
```

This completes the task of finding calibration constants for an instrument serial number passed from an HTML document and displaying the results of that search.

Step 5. Saving PHP output in a server-side file.

Now suppose you wish to process data submitted by a user and save it in a server-side file. Document 1.5, below, is an expansion of Document 1.2 which demonstrates how to save the data shown in the output image for Document 1.2 in a file. The few additional lines that are required are shaded.

Document 1.5. (writeCalib.php)

```
<html>
<head>
<title>Get calibrations for water vapor
instrument</title>
</head>
<body>
<?php
// Extract instrument ID from POST data…
     $SN=$_POST["SN"];
     $len=strlen($SN);
// Open WV instrument calibration constant file…
     $inFile = "WVdata.dat";
     $outFile="WVreport.csv";
     $in = fopen($inFile, "r") or
          exit("Can't open file.");
     $out=fopen(
"c:/Documents and Settings/All Users/
Documents/PHPout/"
.$outFile,"a");
// Read one header line…
     $line=fgets($in);
// Search rest of file for SN match…
     $found=1;
     while ((!feof($in)) && ($found == 1)) {
       $line=fgets($in);
       $values=sscanf($line,"%s %f %f %f %f %f");
       list($SN_dat,$A,$B,$C,$beta,$tau)=$values;
       if (strncasecmp($SN_dat,$SN,$len)==0)
         $found=0;
     }
     fclose($in);
     if ($found == 1)
       echo "Couldn't find this instrument.";
     else {
```

```
// Build table of outputs…
echo  "<p><table border='2'>
      <tr><th>Quantity</th><th>Value</th></tr>";
echo  "</td></tr>";
echo  "<tr><td>Instrument ID</td><td>$SN</td></tr>";
echo  "<tr bgcolor='silver'>
      <td colspan='2'>Calibration Constants</td></tr>";
echo  "<tr><td>A</td><td>$A</td></tr>";
echo  "<tr><td>B</td><td>$B</td></tr>";
echo  "<tr><td>C</td><td>$C</td></tr>";
echo  "<tr><td>&tau;</td><td>$tau</td></tr>";
echo  "<tr><td>&beta;</td><td>$beta</td></tr>";
echo  "</table>";
          }
fprintf($out,"Data have been reported for:
%s,%f,%f,%f,%f,%f\n",$SN,$A,$B,$C,$tau,$beta);
fclose($out);
?>
</body>
</html>
```

The first two highlighted lines create an output file with a .csv extension.

```
$outFile="WVreport.csv";
```

```
$out=fopen(
"c:/Documents and Settings/All Users/
Documents/PHPout/"
.$outFile,"a");
```

You can use any **file name extension** you like—PHP doesn't care—but .csv is the standard extension for "comma-delimited" files that can be opened in Excel and other spreadsheet application. PHP *does* care where you try to store this file. Files can be written only to locations on a server with appropriate write permissions. On my Windows computer, the file is written into a directory that is shared by all users (as opposed to just the administrator, for example), in a separate subfolder, PHPout. In particular, permission is *not* given to write into a file stored in the C:/Inetpub/wwwroot folder of my computer. This folder contains all my PHP applications, and its contents need to be protected in the same way as a remote server needs protection. You will need to choose a folder that is appropriate for your server.

The second of the highlighted lines assigns a file handle, $out, and opens a file with "a" (**append**) permission. This means that if the file already exists, new data will be appended to the end of the file. If the file doesn't already exist, it will be created.

The other common permission is "w" (**write-only**) permission. If the file doesn't already exist, it will be created. If it already exists, all the existing contents of the file will be destroyed and only the results from executing this PHP script will be written in the file. Basically, "write" permission wipes the slate clean each time an output file is created. This may or may not be what you intend, so be careful! In this case, the desired result is to append new data to the end of existing data, not to start over again with an empty file, every time the PHP application executes.

The second group of highlighted lines

```
fprintf($out,"Data have been reported for:
%s,%f,%f,%f,%f,%f\n",$SN,$A,$B,$C,$tau,$beta);
fclose($out);
```

uses a format specifier string to write some text and the values of variables to the output file. The output format specifiers mirror the specifiers used to read these values from the input data file and should match the data type of the information that is going to be written. The commas in the specifier string are written into the output file, which looks like this after one call to writeCalib.php:

```
Data have been reported for:
WV2-157,0.911,0.468,0.2,0.1,0.65
```

It is important to understand that the output created by this PHP application is not in any sense a special "PHP file." It is simply a text file that can be used by other software and applications, including spreadsheets and even other PHP scripts, as needed. In particular, the data in this file are comma-delimited, which means they can be opened directly into a spreadsheet.

As noted previously, the kinds of file access tasks described in this chapter are impossible with JavaScript alone. Of course, there are many PHP language details to be explored, but these are just details compared to the conceptual framework. The examples given here show that by passing values to a PHP document, you can use those values to initiate processing that takes place on a remote server, including accessing existing data and creating new data that can be stored permanently on that server. Many Web programmers use PHP primarily to access databases or validate the contents of submitted online forms. From a science and engineering applications perspective, this will not be the primary use. With PHP's capabilities, and without knowing anything about formal database structure, you can send information from a client-side application—input values for a calculation, for example—to a PHP application and permanently store that information in whatever format you desire, along with results of operations performed on those data, including operations that require access to other information stored on the server. These capabilities vastly expand the range of online applications beyond those that can be carried out with JavaScript alone.

1.2 More Examples

1.2.1 Solving the Quadratic Equation

> For the quadratic equation $ax^2 + bx + c = 0$,
>
> find the real roots:
>
> $r_1 = [-b + (b^2 - 4ac)^{1/2}]/(2a)$ $r_2 = [-b - (b^2 - 4ac)^{1/2}]/(2a)$
>
> The "a" coefficient must not be 0. If the discriminant
>
> $b^2 - 4ac = 0$, there is only one root. If the discriminant is less than 0, there are no real roots.

This problem can be solved easily just with JavaScript, but it provides another example of passing HTML form field values to a PHP application.

Document 1.6a (`quadrat.htm`)

```
<head>
<title>Solving the Quadratic Equation</title>
</head>
<body>
<form method="post" action="quadrat.php">
Enter coefficients for ax<sup>2</sup> + bx + c = 0:
<br />
a = <input type="text" value="1" name="a" />
  (must not be 0) <br />
```

```
b = <input type="text" value="2" name="b" /><br />
c = <input type="text" value="-8" name="c" /><br /><br />
<input type="submit" value="click to get roots..." />
</form>
</body>
</html>
```

Enter coefficients for $ax^2 + bx + c = 0$:

$a =$ `1`   (must not be 0)

$b =$ `2`

$c =$ `-8`

click to get roots...

Document 1.6b (quadrat.php)

```php
<?php
$a = $_POST["a"];
$b = $_POST["b"];
$c = $_POST["c"];
$d = $b*$b - 4.*$a*$c;
if ($d == 0) {
   $r1 = $b/(2.*$a);
   $r2 = "undefined";
}
else if ($d < 0) {
   $r1 = "undefined";
   $r2 = "undefined";
}
else {
   $r1 = (-$b + sqrt($b*$b - 4.*$a*$c))/2./$a;;
   $r2 = (-$b - sqrt($b*$b - 4.*$a*$c))/2./$a;;
}
echo "r1 = " . $r1 . ", r2 = " . $r2;
?>
```

$r1 = 2, r2 = -4$

If the coefficient c is changed from -8 to 8, the equation has no real roots:

$r1 =$ undefined, $r2 =$ undefined

Note that in this example, the PHP variable names are the same as the form field names in the corresponding HTML document. These are reasonable names for coefficients of a quadratic equation, but they could be given any other names, such as $p, $q, and $r, if there were some reason to do that. This PHP document needs to "know" the field names by which these values were identified in the calling HTML document, because those names must be available to $_POST[]. A different approach will be illustrated in the examples at the end of Chapter 3.

1.2.2 Preventing Multiple Submissions from a Form

> Create an HTML document that allows a user to enter some meteorological observations. Pass these observations to a PHP document and append them to a file of observations. Take some steps to prevent a user from submitting the same set of observations more than once.

A typical problem when using a form to send data to a remote server is that it is too easy to submit the same data multiple times just by clicking repeatedly on the "submit" button. There are many approaches to this problem, depending on the application and how hard you wish to make it to change data and resubmit it as a new entry. Document 1.7, below, takes a somewhat relaxed approach based on the assumption that a user may legitimately wish to submit several sets of data on the same "visit" to the form, but should be prevented from sending the same data more than once. Hence, the user is forced to reset the form before the "submit" button will work again.

Note that this is a JavaScript solution, having nothing to do with PHP in the sense that the PHP application doesn't check to see if data that are submitted already exist in the "append" file. It is certainly possible, within the PHP application, to prevent duplicate data from being appended to the file.

Document 1.7a (WeatherReport.htm)

```
<html>
<head>
<title>Weather Report</title>
<script language="javascript" type="text/javascript" >
  var alreadySubmitted = false;
  function submitForm ( )
  {
    if (alreadySubmitted)
```

```
      {
        alert("Data already submitted. Click on 'Reset Button'
and start over." );
        return false;
      }
      else
      {
        alreadySubmitted = true;
        return true;
      }
   }
</script>
</head>
<body>
<h2>Report weather observations</h2>
<form method="post" action="WeatherReport.php"
     onSubmit="return submitForm(this.form);" >
  Date (mm/dd/yyyy) : <input type="text" name="date"
     value="09/23/2007" /><br />
  Time (UT hh:mm:ss): <input type="text" name="time"
     value="17:00:00" /><br />
  Air temperature (deg C):<input type="text" name="T"
     value="23" /><br />
  Barometric pressure (millibar): <input type="text"
     name="BP" value="1010" /><br />
  Cloud cover (octas 0-8): <input type="text" name="octas"
     value="7" /><br />
  Precipitation today (total mm): <input type="text"
     name="precip" value="2.3" /><br />
  <input type="submit" name="PushButton"
     value="Click to submit..." /><br />
  <input type="reset" value="Reset Button"
     onClick="alreadySubmitted=false;"/>
</body>
</html>
```

The comma-delimited text file that will contain the reported data is created ahead of time, starting with just a header line:

```
Date,Time,T,BP,Octas,Precipitation
```

Document 1.7b (WeatherReport.php)

```php
<?php
  $date=$_POST["date"];
  $time=$_POST["time"];
  $T=$_POST["T"];
  $BP=$_POST["BP"];
  $octas=$_POST["octas"];
  $precip=$_POST["precip"];
```

```
echo "You have reported:<br />" .
    "date:" . $date . "<br />" .
    "time: " . $time . "<br />" .
    "BP : " . $BP . "<br />" .
    "octas: " . $octas . "<br />" .
    "precip: " . $precip . "<br />";
$out=fopen("c:/Documents and Settings/
All Users/Documents/phpout/WeatherReport.csv","a");
    fprintf($out,"%s, %s, %.1f, %.2f, %u, %.2f\r\n",
            $date,$time,$T,$BP,$octas,$precip);
    fclose($out);
?>
```

Of course, you will have to change the location of the output file to suit your situation. After two submissions, the `WeatherReport.csv` file might look like this:

```
Date,Time,T,BP,Octas,Precipitation
09/23/2007, 17:00:00, 23.0, 1010.00, 7, 2.30
09/24/2007, 17:10:00, 25.0, 1012.00, 1, 0.00
```

Note that the `fprintf()` format string includes \r\n as a line terminator, so the line breaks will be visible if the file is opened in a Windows text editor such as Notepad, rather than in Excel. Additional format specifiers limit the number of digits associated with the floating-point numbers. See Section 4.4.1 in Chapter 4 for more details about how to use these specifiers.

2 Working with PHP

Chapter 2 introduces the syntax of PHP and shows how to perform calculations. The chapter returns to the problem defined in Chapter 1 and provides a complete PHP-based solution.

2.1 General Structure of PHP Scripts

As previously noted, this book is not a programming tutorial. It assumes you are already familiar with programming concepts such as variables, operators, assignment statements, functions, and loops, and that you have a working knowledge of how these concepts are implemented in JavaScript. Although the file-handling syntax may be unfamiliar to JavaScript programmers, because JavaScript does not have these capabilities, it will not be difficult to learn how to write PHP scripts if you are comfortable with JavaScript. A summary of selected PHP language elements is provided in Chapter 4.

PHP scripts do not appear very different from JavaScript scripts, but there are some important distinctions. For example, PHP scripts *require* a semicolon at the end of each line, but JavaScript scripts do not.

Every PHP variable name must be preceded by a "$" symbol. Variables are not declared ahead of time; there is no equivalent of the JavaScript `var` keyword for declaring a variable without assigning a value. If you use a variable name (for example, `$taxes`) and misspell it later in your script, for example, `$texas`, PHP will not flag this as an error, but your program obviously will no longer give correct results.

PHP scripts can be embedded in HTML documents by using the `<?php … ?>` tag, although that won't be done in the examples found in this book. (There are other ways to embed comments in PHP scripts, including within `<script> … </script>` elements.) `/* … */` is used for block (multiline) comments. Single-line or in-line comments can begin with either `//` or `#`.

PHP supports functions, similar in syntax to JavaScript functions. However, argument passing is simpler because there is no need to distinguish among forms, form fields, and the values of form fields passed

D.R. Brooks, *Introduction to PHP for Scientists and Engineers*,
doi: 10.1007/978-1-84800-237-1_2, © Springer-Verlag London Limited 2008

as inputs. There is no need in PHP for an equivalent of the JavaScript `.value` property, or for the `parseFloat()` or `parseInt()` functions needed to transform form field strings into their corresponding numerical values in calculations or when strings are passed as function arguments.

When JavaScript functions need to return multiple values, one or more forms are passed as input, and outputs are sent to fields in those forms. The same effect can be achieved in PHP by placing multiple values in an array and returning the array. Values returned from a JavaScript function can be used elsewhere in a JavaScript script and then "returned" directly into form fields. PHP values, including output from functions, can also be used within the PHP script, but values cannot be returned to a calling document. Output is returned to a client computer in the form of HTML-formatted output that can be displayed in a browser window.

JavaScript processing is interactive in the sense that you can change inputs and recalculate outputs, sometimes automatically, simply by changing a value in a form field, and sometimes by clicking on a button, all from within the same document. In this sense, PHP works more like old-fashioned command-line "batch processing." If you need another set of outputs, you need to return to the calling document, change the input values in one or more form fields, and send the new values for reprocessing by the PHP script.

From a user's point of view, the biggest difference between JavaScript and PHP is the fact that PHP scripts can read data from and write data to a file on a remote server. This presents potential system security issues, and as a result, institutions that provide web space for authorized users may prohibit the use of *any* server-side languages such as PHP. However, there are no such restrictions with setting up a server on your own computer for your own personal use, although it is certainly possible to overwrite a file that you really couldn't afford to lose!

2.2 Calculations with PHP

As a focus for learning how to do calculations with PHP, return to the problem of calculating column water vapor (precipitable water vapor) based on measurements from a sun photometer, as outlined in Chapter 1. As a first step toward solving this problem, Chapter 1 showed how to pass an instrument serial number to a PHP application so that its calibration constants could be retrieved from a server-side data file.

The next step in this application requires some extensive calculations of the sun's position at a specified location and time on

Earth's surface. Although the details of the algorithms involved are incidental to learning about PHP, the fact that the code is fairly lengthy, involving exponential, logarithmic, and trigonometric functions, will provide many examples of how to do math with PHP.

There are two approaches to doing the calculations required for this problem. The (very short) precipitable water vapor calculation requires as input the output voltages and calibration constants for a specified instrument, and a value for the relative air mass (a dimensionless quantity that has a value of 1 when the sun is overhead, with a solar zenith angle z of $0°$, and which increases as the zenith angle increases, approximately as $1/\cos(z)$). This relative air mass calculation for a particular time and place require some astronomical equations, but they are self-contained and can be done within a JavaScript script. Hence, one option is to calculate the relative air mass in the HTML/JavaScript document and send its value to a PHP application. The other option is to send just the input values—instrument serial number, measurement location, time, and instrument output voltages—to a PHP application, which will then do *all* the required calculations.

There is no compelling reason why one option is better than the other, unless it is considered important to prevent the user from seeing the actual code required to perform the solar position and relative air mass calculations. The justification for choosing the second option here is to take advantage of the opportunity to learn a great deal about how to use the PHP language.

As a starting point, Document 2.1 below is a complete HTML/JavaScript application that calculates precipitable water vapor based on the assumption that the user already knows the calibration constants for the instrument used to collect the data. All the equations for calculating solar position are incorporated into this document, so all that will be required later is to translate them into PHP. If you haven't done these kinds of math-oriented calculations in JavaScript before, as will be true for many JavaScript programmers, it would be an excellent idea to study this document in detail!

Document 2.1. (PWcalc2.htm)

```
<html>
<head>
<title>WV calculations for calibrated
instrument</title>
<script language="javascript">
        document.write("This document last modified on
```

```javascript
"+document.lastModified+".");
</script>
<script language="javascript">
function getSunpos(m,d,y,hour,minute,second,Lat,Lon) {
with (Math) {
// Explicit type conversions to make sure inputs are
treated like numbers, not strings.
  m=parseInt(m,10); d=parseInt(d,10);
  y=parseInt(y,10);
  hour=parseFloat(hour); minute=parseFloat(minute);
second=parseFloat(second);
  Lat=parseFloat(Lat); Lon=parseFloat(Lon);
// Julian date
  var temp=ceil((m-14)/12);
//This number is always <=0.
  var JD = d - 32075 + floor(1461*(y+4800+temp)/4)
    +floor(367*(m-2-temp*12)/12)
// m-2-temp*12 is always > 0.
    -floor(3*(floor((y+4900+temp)/100))/4);
  JD =JD-0.5+hour/24.+minute/1440.+second/86400.;
// Solar position, ecliptic coordinates
  var dr=PI/180.;
  var T=(JD-2451545.0)/36525.0;
  var L0=280.46645+36000.76983*T+0.0003032*T*T;
  var M=357.52910+35999.05030*T-0.0001559*T*T-
0.00000048*T*T*T;
  var M_rad=M*dr;
  var e=0.016708617-0.000042037*T-0.0000001236*T*T;
  var C=(1.914600-0.004817*T-0.000014*T*T)*sin(M_rad)
    +(0.019993-
0.000101*T)*sin(2.*M_rad)+0.000290*sin(3.*M_rad);
  var L_save=(L0+C)/360.;
  if (L_save < 0.) var L_true=(L0+C)-
ceil(L_save)*360.;
  else var L_true=(L0+C)-floor(L_save)*360.;

  if (L_true < 0.) L_true+=360.;
  var f=M_rad+C*dr;
  var R =1.000001018*(1.-e*e)/(1.+e*cos(f));
// Sidereal time
  var Sidereal_time=280.46061837+360.98564736629*(JD-
2451545.)+0.0003879*T*T-T*T*T/38710000.;
  S_save=Sidereal_time/360.;
  if (S_save < 0.) Sidereal_time=Sidereal_time-
ceil(S_save)*360.;
  else Sidereal_time=Sidereal_time-floor(S_save)*360.;
```

```
  if (Sidereal_time < 0.) Sidereal_time+=360.;
// Obliquity
  var obliquity=23.+26./60.+21.448/3600.-
46.8150/3600.*T-0.00059/3600.*T*T
+0.001813/3600.*T*T*T;
// Ecliptic to equatorial
  var right_Ascension =
atan2(sin(L_true*dr)*cos(obliquity*dr),cos(L_true*dr))
;
  var declination =
asin(sin(obliquity*dr)*sin(L_true*dr));
  var Hour_Angle = Sidereal_time + Lon -
right_Ascension/dr;  // Don't know why!!
  var
Elev=asin(sin(Lat*dr)*sin(declination)+cos(Lat*dr)*cos
(declination)*cos(Hour_Angle*dr));
/* relative air mass from Andrew T. Young, Air mass
and refraction (Eq. 5),
 Appl. Opt., 33, 6, 1108-1110 (1994) */
  var cosz=cos(PI/2.-Elev);
  } // End with (Math) …
  var
airm=(1.002432*cosz*cosz+0.148386*cosz+0.0096467)/(cos
z*cosz*cosz+0.149864*cosz*cosz+0.0102963*cosz+0.000303
978);
  return airm;
}
function get_PW(IR1,IR2,A,B,C,beta,tau,airm,p) {
/* NOTE:
   1. Station pressure may be included in these
calculations in the future.
   2. No addition operations in these calculations,
so explicit string conversions to numbers not
required.
*/
    var x = C*airm*tau - (Math.log(IR2/IR1)-A)/B;
    var PW = Math.pow(x,1./beta)/airm;
    return Math.round(PW*1000.)/1000.;
}
</script>
</head>
<body bgcolor="white">
<h2>Calculations for Total Precipitable Water Vapor
(PW)</h2>
<p>
```

```html
<form>
<table border="2">
  <tr bgcolor="silver"><td colspan="4">
    <b>Location:</b></td></tr>
    <td>longitude (decimal degrees): </td>
    <td> <input type="text" name="lon" value="-75.188"
size="8"> </td>
    <td>latitude (decimal degrees): </td><td>
      <input type="text" name="lat"
      value="39.955" size="8"></td>
  </tr>
  <tr bgcolor="silver"><td colspan="4">
    <b>Calibration constants:</b></td>
  </tr>
  <tr><td><b>A (you <u><i>must</i></u> provide a
value)</b>, B, and C:</td>
    <td><input type="text" name="A" value=""
size="8"></td>
    <td><input type="text" name="B" value="0.468"
size="8"></td>
    <td><input type="text" name="C" value="0.2"
size="8"></td>
    <tr><td colspan="2">&beta; and &tau;:</td>
    <td><input type="text" name="beta" value="0.65"
size="4"></td>
    <td><input type="text" name="tau" value="0.10"
size="4"></td>
  </tr>
  <tr bgcolor="silver"><td
colspan="4"><b>Date:</b></td>
  </tr>
  <tr><td>mm/dd/yyyy</td>
    <td><input type="text" name="mon" value="4"
size="3"></td>
    <td><input type="text" name="day" value="5"
size="3"></td>
    <td><input type="text" name="yr" value="2007"
size="5"></td>
  </tr>
  <tr bgcolor="silver"><td
colspan="4"><b>Time:</b></td></tr>
    <tr><td>hh:mm:ss (<b><i><u>must</u></i> be Universal
Time</b>)</td>
    <td><input type="text" name="hr" value="14"
size="3"></td>
    <td><input type="text" name="min" value="33"
size="3"></td>
```

```
    <td><input type="text" name="sec" value="15"
size="3"></td>
  </tr>
  <tr><td colspan="2" bgcolor="silver">
    <b>Station pressure (mbar, not currently used in
calculation):</b></td>
    <td colspan="2"><input type="text" name="p"
value="1013"size="7"></td>
  </tr>
  <tr bgcolor="silver"><td colspan="4"><b>Instrument
voltages:</b></td></tr>
  <tr>
    <td>IR1</td>
    <td><input type="text" name="IR1" value="0.742"
size="5"> </td>
    <td>IR1<sub>dark</sub></td>
    <td><input type="text" name="IR1_dark"
value="0.003" size="5"></td>
  </tr>
  <tr>
    <td>IR2</td>
    <td><input type="text" name="IR2" value="0.963"
size="5"></td>
    <td>IR2<sub>dark</sub></td>
    <td><input type="text" name="IR2_dark"
value="0.004" size="5"></td>
  </tr>
</table>
<input type="button" value="Click here to calculate
relative air mass and PW"
onclick="
// Get relative air mass…
airm.value=getSunpos(this.form.mon.value,
  this.form.day.value,this.form.yr.value,
  this.form.hr.value,this.form.min.value,
  this.form.sec.value,
  this.form.lat.value,this.form.lon.value);
// then PW…
PW.value=get_PW(this.form.IR1.value-
this.form.IR1_dark.value,
  this.form.IR2.value-this.form.IR2_dark.value,
  this.form.A.value,this.form.B.value,
  this.form.C.value,this.form.beta.value,
  this.form.tau.value,
  this.form.airm.value,this.form.p.value);
airm.value=Math.round(airm.value*10000.)/10000.;">
```

```
<br />
Relative air mass: <input type="text" name="airm"
value="0"size="7">
<br />
Overhead precipitable water vapor (cm H<sub>2</sub>O):
<input type="text" name="PW" value="0" size="7">
</form>
</body>
</html>
```

Location:			
longitude (decimal degrees):	-75.188	latitude (decimal degrees):	39.955
Calibration constants:			
A (you *must* provide a value), B, and C:	1.123	0.468	0.2
β and τ:		0.65	0.10
Date:			
mm/dd/yyyy	4	5	2007
Time:			
hh:mm:ss (*must* be Universal Time)	14	33	15
Station pressure (mbar, not currently used in calculation):	1013		
Instrument voltages:			
IR1	0.742	IR1$_{dark}$	0.003
IR2	0.963	IR2$_{dark}$	0.004

Click here to calculate relative air mass and PW

Relative air mass: 1.4852

Overhead precipitable water vapor (cm H$_2$O): 1.767

To start the transition from JavaScript to PHP, Document 2.2 below is an HTML document that will pass instrument and measurement data to a PHP application:

Document 2.2. (PWcalc3.htm)

```
<html>
<head>
<title>WV calculations for calibrated
instrument</title>
<script language="javascript">
 document.write("This document last modified on "
   +document.lastModified+".");
</script>
```

```
</head>
<body bgcolor="white">
<h2>Calculations for Total Precipitable Water Vapor
(PW)</h2>
<p>
<form method="post" action="PWcalc3.php">
<table border="2">
  <tr bgcolor="silver"><td
      colspan="4"><b>Location:</b></td></tr>
  <td>longitude (decimal degrees): </td>
  <td> <input type="text" name="lon" value="-75.188"
     size="8"> </td>
  <td>latitude (decimal degrees): </td>
  <td><input type="text" name="lat" value="39.955"
     size="8"></td></tr>
  <tr bgcolor="silver"><td colspan="4">
    <b>Instrument Serial Number:</b></td></tr>
  <tr><td colspan="4"><input type="text" name="SN"
     value="WV2-117" /></td></tr>
  <tr bgcolor="silver">
    <td colspan="4"><b>Date:</b></td></tr>
  <tr><td>mm/dd/yyyy</td>
  <td><input type="text" name="mon" value="4"
     size="3"></td>
  <td><input type="text" name="day" value="5"
     size="3"></td>
  <td><input type="text" name="yr" value="2005"
     size="5"></td></tr>
  <tr bgcolor="silver"><td
colspan="4"><b>Time:</b></td>
  <tr><td>hh:mm:ss (<b><i><u>must</u></i> be Universal
     Time</b>)</td>
  <td><input type="text" name="hr" value="14"
     size="3"></td>
  <td><input type="text" name="min" value="33"
     size="3"></td>
  <td><input type="text" name="sec" value="15"
     size="3"></td></tr>
  <tr><td colspan="2" bgcolor="silver"><b>Station
     pressure (mbar, not currently used in
     calculation):</b></td>
  <td colspan="2"><input type="text" name="p"
     value="1013" size="7"></td></tr>
  <tr bgcolor="silver"><td colspan="4">
    <b>Instrument voltages:</b></td></tr>
  <tr>
```

```
<td>IR1</td>
<td><input type="text" name="IR1" value="0.742"
   size="5"> </td>
<td>IR1<sub>dark</sub></td>
<td><input type="text" name="IR1_dark" value="0.003"
   size="5"></td>
</tr>
<tr>
<td>IR2</td>
<td><input type="text" name="IR2" value="0.963"
   size="5"></td>
<td>IR2<sub>dark</sub></td>
<td><input type="text" name="IR2_dark" value="0.004"
   size="5"></td>
</tr>
</table>
<input type="submit"
   value="Click here to calculate PW..." />
</form>
</body>
</html>
```

This document last modified on 01/16/2008 10:39:39.

Calculations for Total Precipitable Water Vapor (PW)

Location:			
longitude (decimal degrees):	-75.188	latitude (decimal degrees):	39.955
Instrument Serial Number:			
WV2-117			
Date:			
mm/dd/yyyy	4	5	2005
Time:			
hh:mm:ss (*must* be Universal Time)	14	33	15
Station pressure (mbar, not currently used in calculation):	1013		
Instrument voltages:			
IR1	0.742	IR1$_{dark}$	0.003
IR2	0.963	IR2$_{dark}$	0.004

Click here to calculate PW...

The output looks similar to the output for Document 2.1, except that the PW calculations are replaced with a "submit" button that will pass values to a PHP document. The companion Document 2.3, below, is a PHP application that will accept input values from Document 2.2, look up calibration constants, and then calculate PW. It incorporates the previous code from Document 1.2, which accepted an input value from an HTML document and used that value to extract calibration constants from a data file stored on a server.

Document 2.3. (PWcalc3.php)

```php
<html>
<title>WV calculations for calibrated instrument
</title>
<?php
function getJD($m,$d,$y,$hour,$minute,$second) {
// Julian date
  $temp=ceil(($m-14)/12);//This number is always <= 0.
  $JD = $d - 32075 + floor(1461*($y+4800+$temp)/4)
    +floor(367*($m-2-$temp*12)/12) // m-2-temp*12 is
always > 0.
     -floor(3*(floor(($y+4900+$temp)/100))/4);
  $JD =$JD-0.5+$hour/24.+$minute/1440.+$second/86400.;
  return $JD;
}
function
getSunpos($m,$d,$y,$hour,$minute,$second,$Lat,$Lon) {
// Retrieve Julian date
  $JD=getJD($m,$d,$y,$hour,$minute,$second);
// Solar position, ecliptic coordinates
  $dr=pi()/180.;
  $T=($JD-2451545.0)/36525.0;
  $L0=280.46645+36000.76983*$T+0.0003032*$T*$T;
  $M=357.52910+35999.05030*$T-0.0001559*$T*$T-
0.00000048*$T*$T*$T;
  $M_rad=$M*$dr;
  $e=0.016708617-0.000042037*$T-0.0000001236*$T*$T;
  $C=(1.914600-0.004817*$T-0.000014*$T*$T)*sin($M_rad)
    +(0.019993-
0.000101*$T)*sin(2.*$M_rad)+0.000290*sin(3.*$M_rad);
// Replacement code for L_true=fmod(L0+c,360.)
  $L_save=($L0+$C)/360.;
  if ($L_save < 0.) $L_true=($L0+$C)-
ceil($L_save)*360.;
  else $L_true=($L0+$C)-floor($L_save)*360.;
  if ($L_true < 0.) $L_true+=360.;
  $f=$M_rad+$C*$dr;
  $R =1.000001018*(1.-$e*$e)/(1.+$e*cos($f));
// Sidereal time
  $Sidereal_time=280.46061837+360.98564736629*($JD-
```

```php
2451545.)+0.0003879*$T*$T-$T*$T*$T/38710000.;
// Replacement code for Sidereal=fmod(Sidereal,360.)
   $S_save=$Sidereal_time/360.;
   if ($S_save < 0.) $Sidereal_time=$Sidereal_time-
ceil($S_save)*360.;
   else $Sidereal_time=$Sidereal_time-
floor($S_save)*360.;
   if ($Sidereal_time < 0.) $Sidereal_time+=360.;
// Obliquity
   $obliquity=23.+26./60.+21.448/3600.-
46.8150/3600.*$T-0.00059/3600.*$T*$T
+0.001813/3600.*$T*$T*$T;
// Ecliptic to equatorial
   $right_Ascension =
atan2(sin($L_true*$dr)*cos($obliquity*$dr),cos($L_true
*$dr));
   $declination =
asin(sin($obliquity*$dr)*sin($L_true*$dr));
   $Hour_Angle = $Sidereal_time + $Lon -
$right_Ascension/$dr;
$elev=asin(sin($Lat*$dr)*sin($declination)+cos($Lat*$d
r)*cos($declination)*cos($Hour_Angle*$dr));
/* relative air mass from Andrew T. Young, Air mass
and refraction (Eq. 5),
 Appl. Opt., 33, 6, 1108-1110 (1994) */
   $cosz=cos(pi()/2.-$elev);
//  echo $cosz;
$airm=(1.002432*$cosz*$cosz+0.148386*$cosz+0.0096467)/
($cosz*$cosz*$cosz+
0.149864*$cosz*$cosz+0.0102963*$cosz+0.000303978);
   return $airm;
}
?>
</head>
<body bgcolor="white">
<?php
echo "<h2>Calculations for Total Precipitable Water
Vapor (PW)</h2>";
$m=getSunpos($_POST["mon"],$_POST["day"],$_POST["yr"],
$_POST["hr"],$_POST["min"],$_POST["sec"],
$_POST["lat"],$_POST["lon"]);
$IR1=$_POST["IR1"]-$_POST["IR1_dark"];
$IR2=$_POST["IR2"]-$_POST["IR2_dark"];
      $A=$_POST["A"];
      $B=$_POST["B"];
      $C=$_POST["C"];
      $beta=$_POST["beta"];
```

```php
      $tau=$_POST["tau"];
      $SN=$_POST["SN"];
      $len=strlen($SN);
// Open WV instrument calibration constant file...
      $inFile = "WVdata.dat";
      $in = fopen($inFile, 'r') or die("Can't open
file");
// Read one header line...
      $line=fgets($in);
// Search rest of file for SN match...
      $found=1;
      while ((!feof($in)) && ($found == 1)) {
        $line=fgets($in);
        $values=sscanf($line,"%s %f %f %f %f %f\n");
        list($SN_dat,$A,$B,$C,$beta,$tau)=$values;
        if (strncasecmp($SN_dat,$SN,$len)==0) {
          $found=0;
        }
      }
    fclose($in);
// Build table of outputs...
echo "<p><table
border='2'><tr><th>Input</th><th>Value</th></tr>";

echo "</td></tr>";
echo "<tr><td>Instrument SN</td><td>$SN</td></tr>";
echo "<tr bgcolor='silver'><td colspan='2'>Calibration
Constants</td></tr>";
echo "<tr><td>A</td><td>$A</td></tr>";
echo "<tr><td>B</td><td>$B</td></tr>";
echo "<tr><td>C</td><td>$C</td></tr>";
echo "<tr><td>&tau;</td><td>$tau</td></tr>";
echo "<tr><td>&beta;</td><td>$beta</td></tr>";
echo "<tr bgcolor='silver'>
  <td colspan='2'>Measurements</td></tr>";
echo "<tr><td>IR1 (sunlight -
dark)</td><td>$IR1</td></tr>";
echo "<tr><td>IR2 (sunlight -
dark)</td><td>$IR2</td></tr>";
echo "</table></p>";
$x = $C*$m*$tau - (log($IR2/$IR1)-$A)/$B;
$PW = pow($x,1./$beta)/$m;
echo "<p><table border='2'><tr><th>Output</th>
  <th>Value</th></tr>";
echo "<tr><td>relative air mass</td><td>";
echo round($m,4);
```

```
echo "</td></tr>";
echo "<tr><td>PW, cm H<sub>2</sub>0</td><td>";
echo round($PW,4);
echo "</table></p>";
?>
</body>
</html>
```

Calculations for Total Precipitable Water Vapor (PW)

Input	Value
Instrument SN	WV2-117
Calibration Constants	
A	0.9793
B	0.4687
C	0.2
τ	0.1
β	0.65
Measurements	
IR1 (sunlight - dark)	0.739
IR2 (sunlight - dark)	0.959

Output	Value
relative air mass	1.4806
PW, cm H_2O	1.3427

Needless to say, Document 2.3 deserves close attention and line-by-line comparison with Document 2.1, because it contains a great deal of information about using PHP in your own applications. Here are some general observations about similarities and differences between JavaScript and PHP.

1. The syntax of writing expressions and statements is essentially the same.

2. The syntax for user-defined functions is essentially the same.

3. In PHP, mathematical calculations are carried out with built-in functions, such as $\sin(x)$, rather than the "methods" such as

`Math.sin(x)` used in JavaScript. As a practical matter, distinctions between a "function" and a "method" at the conceptual and language design level don't matter in these applications.

4. Values passed to a PHP application do not need to be converted explicitly from strings to numbers. That is, there is no PHP equivalent to JavaScript's `parseFloat()` method. Recall that, in JavaScript, the result of adding two "numbers" passed from a form field would be a string that contained the concatenation of the two "numbers," interpreted as though they were strings of characters. In PHP, the concatenation operator is a period, not a "+" sign, so there is no chance for confusion. As a practical matter, it appears to be safe to conclude that the operator used on a value determines how its operands will be treated; the statement `$C = $A + $B;` interprets `$A` and `$B` as two numbers because this is the only interpretation that makes sense for the addition operator.

Considering that PHP and JavaScript are two different languages, the translation from JavaScript to PHP is remarkably easy. The explicit conversions of form field value from strings to numbers that are required in JavaScript is replaced by PHP code that assigns variables based on the values passed to the `$_POST[]` array in Document 2.3:

```
$m=getSunpos($_POST["mon"],$_POST["day"],$_POST["yr"],
$_POST["hr"],
$_POST["min"],$_POST["sec"],$_POST["lat"],
$_POST["lon"]);
//$m=$_POST["airm"];
$IR1=$_POST["IR1"]-$_POST["IR1_dark"];
$IR2=$_POST["IR2"]-$_POST["IR2_dark"];
$A=$_POST["A"];
$B=$_POST["B"];
$C=$_POST["C"];
$beta=$_POST["beta"];
$tau=$_POST["tau"];
$SN=$_POST["SN"];
$len=strlen($SN);
```

`$_POST` is the PHP-generated array containing all the form field values passed to the application. For the most part, the new variable names created in Document 2.3 are the same as the names in the form fields from the calling HTML/JavaScript application. However, this doesn't have to be true. The statements

```php
$IR1=$_POST["IR1"]-$_POST["IR1_dark"];
$IR2=$_POST["IR2"]-$_POST["IR2_dark"];
```

take advantage of the fact that the voltage from each channel required for the precipitable water vapor calculation

```php
$x = $C*$m*$tau - (log($IR2/$IR1)-$A)/$B;
$PW = pow($x,1./$beta)/$m;
```

is the voltage reported when the instrument is pointed at the sun, minus the "dark" voltage produced by the instrument's electronics. So, $IR1 and $IR2 are defined as the difference between the sunlight and dark voltages for each channel, as posted from the calling document.

When converting JavaScript to PHP it is critical to remember to add a $ character to the beginning of all variable names. If you forget this character, PHP won't produce an error message, but your program will certainly not work!

The PHP syntax for creating user-defined functions that return a single value is just like JavaScript:

```php
function
getSunpos($m,$d,$y,$hour,$minute,$second,$Lat,$Lon) {
    ...

$airm=(1.002432*$cosz*$cosz+0.148386*$cosz+0.0096467)/
($cosz*$cosz*$cosz+
   0.149864*$cosz*$cosz+0.0102963*$cosz+0.000303978);
  return $airm;
}
```

A PHP function for rounding numbers to a specified number of digits to the right of the decimal point can be used to limit the number of digits that would otherwise be displayed in an output string. For calculations based on the physical world, many of the digits displayed by default are meaningless. This statement rounds off the relative air mass to four digits:

```php
echo round($m,4);
```

The round() function will not retain significant digits when they are 0. That is, round(5.444,4) displays 5.444 rather than 5.4440. (You can gain more control over output by using other output functions such as printf().)

With PHP functions, it is the programmer's responsibility to ensure that input arguments are used appropriately because no syntax distinction is made between, for example, arguments intended to be used as character strings and those intended to be used as numbers. On the other hand, it is not necessary to worry about arguments that "look" like numbers being treated as strings, as could be the case in JavaScript.

The statement below from `function getSunpos()` in Document 2.3 demonstrates that a user-defined PHP function can be called from inside another PHP function, as expected:

```
$JD=getJD($m,$d,$y,$hour,$minute,$second);
```

One important topic not addressed by Document 2.3 is how to return multiple values from a PHP function. Recall that in JavaScript, this is accomplished by passing the name of a form (equivalent to passing its location in memory). The values of multiple fields defined in that form can then be changed, with no `return` statement.

In PHP, multiple values are returned by using the `array()` **constructor** to create an array of the values you wish to return from a function and then `return` that array. (Arrays will be discussed in more detail in Chapter 3.) The `list()` construct can then be used to extract the values from that array. Document 2.4 shows how to do this.

Document 2.4. (`circleStuff.php`)

```php
<?php
/* function CIRCLESTUFF($r) {...}
   will also work because PHP function names are case-
   insensitive!
*/
  function CircleStuff($r) {
        $area=M_PI*$r*$r;
        $circumference=2.*M_PI*$r;
/* However, this won't work:
    return array($AREA,$circumference);
   because variable names are case-sensitive.
*/
    return array($area,$circumference);
  }
    list($area,$circumference) = CircleStuff(3.);
    echo $area . ", " . $circumference;
?>
```

The `echo` statement displays the following:

```
28.274333882308, 18.849555921539
```

(This result begs for the application of the `round()` function!

It is a peculiarity of PHP that function names are case-*in*sensitive. Thus, `function CircleStuff()` and `function CIRCLESTUFF()` will both work in this example. Because variable names are case-sensitive, and because great care is generally required in matching cases in all other aspects of programming, it makes little sense to take advantage of this PHP "feature."

2.3 More about PHP File Input/Output

Consider the following problem:

A text file contains wind speed data:

```
1 1991 31
 3.2, 0.4, 3.8, 4.5, 3.3, 1.9, 1.6, 3.7, 0.8, 2.3,
2.8, 2.4, 2.5, 3.2, 4.1, 3.9, 5.0, 4.4, 4.4, 5.5, 3.0,
3.7, 2.2, 2.0
 2.6, 2.8, 2.3, 2.3, 1.2, 2.4, 3.1, 4.0, 3.6, 2.9,
6.0, 4.4, 0.8, 3.8, 3.5, 4.5, 2.7, 3.4, 6.6, 5.2, 1.6,
1.2, 2.3, 2.4
...
2 1991 28
 4.6, 5.9, 3.1, 3.2, 4.5, 4.4, 3.9, 4.4, 7.5,
8.4,10.2, 9.2, 8.1, 6.3, 3.1, 3.5, 2.2, 1.4, 0.4, 4.2,
5.4, 4.0, 2.9, 1.7
 2.5, 2.3, 2.1, 1.5, 2.3, 4.1, 5.3, 6.0, 6.0,
9.7,11.3,12.7,13.0,13.0,11.6, 9.9, 9.6, 8.7, 5.4, 5.1,
5.3, 5.6, 4.4, 4.2
...
```

The three numbers in the first line of the file are the month, year, and number of days in the month. Then, for each day in the month, 24 hourly wind speeds are given (in units of miles per hour), separated by commas. Each set of 24 hourly values is on the same line of text in the file, even though each of those lines occupies three lines as displayed here. This pattern is repeated for all 12 months. Missing data are represented by a value of -1.

Write a PHP script that will read this file and count the number of missing values for each month. It should display as

> output the number of each month (1–12) the year, and the number
> of missing values for that month. Write the results into a file and
> save it.

The calculations required for this problem are simple, but reading
the data file correctly requires some care. Document 2.5 shows the code
for solving this problem.

Document 2.5. (windspd.php)

```php
<?php
$inFile="windspd.dat";
$outFile="windspd.out";
$in = fopen($inFile, "r") or die("Can't open file.");
$out=fopen("c:/Documents and Settings/
        All Users/Documents/phpout/".$outFile,"w");
while (!feof($in)) {
// Read one month, year, # of days.
  fscanf($in,"%u %u %u",$m,$y,$nDays);
  if (feof($in)) exit;
  echo $m . ', ' . $y . ', ' . $nDays . '<br />';
  $nMissing=0;
  for ($i=1; $i<=$nDays; $i++) {
    $hrly = fscanf($in,"%f,%f,%f,%f,%f,%f,%f,%f,%f,
%f,%f,%f,%f,%f,%f,%f,%f,%f,%f,%f,%f,%f,%f,%f");
    for ($hr=0; $hr<24; $hr++) {
      // echo $hrly[$hr] . ', ';
      if ($hrly[$hr] == -1) $nMissing++;
    }
    // echo $hrly[23] . '<br />';
  }
  echo 'Number of missing hours this month is ' . $nMissing
      .'.<br />';
      fprintf($out,"%u, %u, %u\r\n",$m,$y,$nMissing);
}
echo "All done.<br />"
// fclose($in);
// fclose($out);
?>
```

```
1, 1991, 31
Number of missing hours this month is 22.
2, 1991, 28
Number of missing hours this month is 0.
All done.
```

The input file required by Document 2.5, `windspd.dat`, is stored in the PHP document folder, and the output file is written to the directory set aside for this purpose. The output shown here is for a short version of this file, with data for only two months.

The `feof()` function is used to test for an end-of-file mark that, when found, terminates the program. It is often the case that code to read data from a data file should not assume ahead of time how many values are in the file. Thus, a conditional loop is most often the appropriate approach. In Document 2.5, calls to the `fclose()` function, used in earlier code testing, are commented out because they will not be executed. Execution of the `exit` construct will close all open files.

If additional processing is required after reaching the end of the file, the alternative is to use `break` rather than `exit`:

```
while (!feof($in)) {
// Read one month, year, # of days.
      fscanf($in, "%u %u %u",$m,$y,$nDays);
      if (feof($in)) break;
...
}
echo "All done.<br />";
fclose($in);
fclose($out);
```

Executing a `break` exits the loop and code execution continues starting with the first statement after the loop.

The hourly data are read with `fscanf()`. Especially if you are a C programmer, you might be tempted (I was!) to try reading the 24 hourly wind speed values like this:

```
// PHP code that won't work!
for ($hr = 0; $hr<23; $hr++)
      fscanf($in, "%f,", $hrly[i]);
fscanf($in,"%f",$hrly[23]);
```

This code assumes that `fscanf()` can be used to read values from the file one at a time. However, this won't work in PHP. The `fscanf()` function reads an entire line of text, just as `fgets()` does, regardless of what appears in the format string. The difference is that, without providing specific variable names to be read from the file, `fscanf()` puts values in an array, whereas `fgets()` puts everything in a string that must then be parsed with `sscanf()`. So, you can read the entire 24 hours worth of wind speeds with a single call to `fscanf()`, but you need to write out 24 format specifiers, as shown.[1]

Note these two `echo` statements inside the `while...` loop in Document 2.5:

```
// echo $hrly[$hr] . ', ';
...
// echo $hrly[23] . '<br />';
```

If the `//`'s are removed, all the wind speed values will be displayed. Whenever you are reading a file, it is important to ensure that you are reading the file correctly. The best way to do this is to echo back values from the file. If they all have the expected values, by comparison with the original data file, then you can proceed.

Part of the output for this problem is an output file that summarizes the missing data:

(`windspd.out`)

```
1, 1991, 22
2, 1991, 0
```

The file is opened in write-only mode and the data are written with `fprintf()` in the shaded line in Document 2.5. Note that because this code was written on a Windows computer and the output file will be used on a Windows computer, each line is terminated not just with \n, but with \r\n. This is because Windows text files have both a "linefeed" and a "return" character at the end of each line. Recall Document 1.5, in which only the \n character was used as a line termination. Although it is not obvious, the fact that this file was created as a `.csv` file and opened directly into Excel means that only the line feed \n character was needed. If you open the same file in Notepad, for example, there will be no line breaks.

2.4 Another Example

Write an HTML document that allows a user to select a solid object shape and enter its dimensions and the material from which

---

[1] This is a perfect example of the why the preface warned against assuming that PHP works like C just because the syntax looks the same!

it is made. The choices could be a cube, a rectangular block, a cylinder, or a sphere. You could choose a number of possible materials—air, gold, water, etc. Then call a PHP application that will find the mass of the object by calculating its volume based on the specified shape and the density of the material as retrieved from a data file.

Document 2.6a shows an HTML interface for this problem. The possible shapes and materials are placed in `<select>` lists.

Document 2.6a (getMass.htm)

```
<html>
<head>
<title>Calculate mass</title>
</head>
<body>
<form name="form1" method="post" action="getMass.php">
Enter length: <input type="text" name="L" value="3" /><br />
Enter width: <input type="text" name="W" value="2" /><br />
Enter height: <input type="text" name="H" value="10" /><br />
Enter radius: <input type="text" name="R" value="3" /><br />
<select name="shapes" size="10">
  <option value="cube">cube</option>
  <option value="cylinder">cylinder</option>
  <option value="block">rectangular block</option>
  <option value="sphere">sphere</option>
</select>
<select name="material" size="10">
  <option value="air">air</option>
  <option value="aluminum">aluminum</option>
  <option value="gold">gold</option>
  <option value="oxygen">oxygen</option>
  <option value="silver">silver</option>
  <option value="water">water</option>
</select>
<input type="submit" value="Click to get volume."
<!--
<input type="button" value="click"
onclick="alert(shapes.selectedIndex);
alert(material.options[material.selectedIndex].value); " />
-->
  />
</form>
</body>
</html>
```

Note that the `value` attribute of the `<option>` tag can be, but does not have to be, the same as the text for the option. For the rectangular block shape, `value` is assigned as a single word (`block`), which will look like a single string literal value when it is used later in the PHP application.

In the original version of Document 2.6a, the first line of the input tag line,

```
<input type="submit" value="Click to get volume."
```

was replaced with the shaded lines that are now commented out of the `<input>` tag near the end of Document 2.6a. This code will show which item has been chosen for each `<select>` list. It remains to be seen how this information will be handled by the PHP application.

It is almost never a good idea to try to write an entire JavaScript or PHP application all at once. A much better approach is to proceed step by step, testing your results one step at a time. Once you understand Document 2.6a, it is then worth writing a single-line PHP application that will simply look at what is posted to the application by using the `print_r()` function to display the contents of the `$_POST` array:

```
<?php
  print_r($_POST);
?>
```

This code will display something like this:

```
array ( [L] => 1 [W] => 1 [H] => 1 [R] => 3 [shapes] => cube
[material] => oxygen)
```

Once you are convinced that the inputs are successfully passed to PHP, how should the calculations be done? The obvious first step is to create a data file containing materials and their densities:

(`density.dat`)

```
material density (kg/m^3)
water 1000
aluminum 2700
gold 19300
silver 10500
oxygen 1.429
air 1.2
```

The header line is optional, but it is always a good idea to describe the contents of a data file, including, in this case, the physical units in which the densities should be supplied.

The next step is less obvious. Although it is certainly possible to "hard code" volume calculations for each allowed shape, a more interesting solution is to create a second data file that contains PHP code for calculating the volume of each shape:

(volume.dat)

```
shape volume
cube $L*$L*$L
sphere 4./3.*M_PI*$R*$R*$R
cylinder M_PI*$R*$R*$L
block $L*$W*$H
```

The code string for each allowed shape assumes specific variable names for the dimensions: $L, $W, $H, and $R.

Start building the PHP application like this:

```
<?php
print_r($_POST);
$material=$_POST[material];
$shape=$_POST[shapes];
$L=$_POST[L];
$W=$_POST[W];
$H=$_POST[H];
$R=$_POST[R];
echo "<br />" . $material . ", " . $shape . "<br />";
?>
```

This code will display:

```
array ( [L] => 1 [W] => 1 [H] => 1 [R] => 3 [shapes] => cube
[material] => oxygen )
oxygen, cube
```

and now it is clear that the PHP application is properly receiving the inputs passed from Document 2.6a. (You could also echo the values of $L, $W, $H, and $R if you like.) In Document 2.6a, the fields were given the names L, W, H, and R, but this would not need to be the case. All that is important for the PHP application is to give the variables the same names used in the volume.dat file.

Document 2.6b gives the entire solution to this problem. This code should be written in three sections: first, the definition of the variables as shown above, then the code to search for the material in its data file, and finally the code to do the mass calculation.

Document 2.6b (getMass.php)

```php
<?php
print_r($_POST);
$material=$_POST[material];
$shape=$_POST[shapes];
$L=$_POST[L];
$W=$_POST[W];
$H=$_POST[H];
$R=$_POST[R];
echo "<br />" . $material . ", " . $shape . "<br />";
$materialFile=fopen("density.dat","r");
$shapeFile=fopen("volume.dat","r");
// Read materials file.
$found=false;
$line=fgets($materialFile);
while ((!feof($materialFile)) && (!$found)) {
  $values=fscanf($materialFile,"%s %f",$m,$d);
  if (strcasecmp($material,$m) == 0) {
    echo $material . ", " . $m . ", " . $d . "<br />";
      $found=true;
  }
}
// Read volume file.
$found=false;
$line=fgets($shapeFile);
while ((!feof($shapeFile)) && (!$found)) {
  $values=fscanf($shapeFile,"%s %s",$s,$v);
  if (strcasecmp($shape,$s) == 0) {
    echo $shape . ", " . $v . "<br />";
      $found=true;
  }
}
// Close both data files.
fclose($materialFile);
fclose($shapeFile);
// Calculate mass.
$vv=$v . "*$d";
echo "Result: ".eval("return round($vv,3);")." kg<br />";
?>
```

The next step is to use a conditional loop to look through the material density data file to find a match with the material supplied. In the interests of demonstrating just the essentials, Document 2.6b does not

include code to determine whether the supplied material is included in the
file of materials, but this would not be difficult to do.

The final step is to look through the volume calculation file to
find a match with the shape supplied. Again, the assumption is that the
supplied shape will be represented in the file.

The not-so-obvious and rather clever part of this application is
included in the two shaded lines of code in Document 2.6b:

```
$vv=$v . "*$d";
echo "Result: " . eval("return round($vv,3);") . "<br />";
```

The first of these lines appends `"*$d"` to the volume calculation
string—mass equals volume times density. This string now looks like a
"legal" PHP statement; for example,

```
M_PI*$R*$R*$L*$d
```

(You could `echo` the value of `$vv` if you wanted to see what it contained.)
The next line of code "executes" this statement, using the `eval()`
construct (which looks like a function, but is not), which is similar to the
JavaScript `eval()` global method. The `return` keyword is required to
get back the numerical result, and the `round()` function is applied to the
calculation to remove extraneous digits from the output. The obvious
advantage of this approach is that you can add new materials and shapes
without altering the PHP code, assuming that, at most, four variables—
length, width, height, and radius—will be sufficient to describe all
dimensions needed for the volume calculations. For more complicated
shapes, it might be necessary to add new variables or apply different
interpretations to existing variables.

3 PHP Arrays

Chapter 3 provides an introduction to PHP arrays. The complexity of the PHP array model compared to JavaScript provides many opportunities for creative ways of accessing and manipulating data.

3.1 Array Definition

The ability to organize related information in arrays is very important in science and engineering applications. An array is a programmer-defined data structure that is used to access and manipulate related information. An array has a name, two or more locations (elements) for storing information, and indices or keys for accessing the elements.

PHP supports *dozens* of functions and constructs for manipulating arrays, corresponding to a much more complicated model of how arrays work. Hence, it is not sufficient simply to translate JavaScript array syntax. This chapter will present just the basics of working with PHP arrays.

In JavaScript's conceptual model for arrays, array elements can contain a mixture of data types. Each element is accessed with an integer index. The index of the first element is always 0.

In PHP, arrays are created with the `array()` constructor:

```
$ArrayName = array([(mixed data types)…])
```

where `$ArrayName` is a generic representation of a user-supplied array name. The elements can contain a mixture of data types, just like JavaScript arrays. However, in PHP, each element of an array can have its own user-defined index (key) value:

```
$a = array($key1 => $value1, $key2 => $value2,
           $key3 => $value3,…);
```

The `=>` operator associates a key with its value. The keys can be numbers, characters, or strings. Numerical keys can start with any value, not just 0,

D.R. Brooks, *Introduction to PHP for Scientists and Engineers*,
doi: 10.1007/978-1-84800-237-1_3, © Springer-Verlag London Limited 2008

and they don't even have to be sequential (although they usually are). Document 3.1 shows an example of an array with named keys.

Document 3.1. (`keyedArray.php`)

```php
<?php
// Create an array with user-specified keys...
echo '<br />A keyed array:<br />';
$stuff = array('mine' => 'BMW', 'yours' => 'Lexus',
   'ours' => 'house');
foreach ($stuff as $key => $val) {
  echo '$stuff[' . $key . '] = '. $val . '<br />';
}
?>
```

```
A keyed array:
$stuff[mine]  = BMW
$stuff[yours] = Lexus
$stuff[ours]  = house
```

The names associated with the keys and array elements can be anything you like—they don't have to be `$key` and `$val`, as they are in Document 3.1. For an array with string names for the keys, you need to use a `foreach...` loop, with syntax as shown in the shaded statement. It is the syntax following the `as` keyword that makes the association between a key name and its array element.

The number of elements in an array is given by the `sizeof()` function. Note that a `foreach...` loop does not require or even allow that you specify the length of the array. The syntax that should be familiar to JavaScript programmers, using `sizeof()` and an integer index inside a `for...` loop, won't work with this array because the indices have arbitrary names rather than sequential values:

```php
/* This won't work!
for ($i=0; $i<sizeof($stuff); $i++)
  echo $stuff[$i] . '<br />';
*/
```

This `for...` loop code won't generate an error message—it just won't generate any output.

If the keys specified are the default integer keys starting at 0, then it is straightforward to use a JavaScript-like `for...` loop, as shown in Document 3.2, below. However, it is also possible to use a `for...` loop if the array is created with a starting index other than 0, or if it has consecutive character keys; these possibilities are also illustrated in Document 3.2.

Document 3.2. (`ConsecutiveKeyArray.php`)

```php
<?php
$a = array('david','apple','Xena','Sue');
echo "Using for... loop<br />";
for ($i=0; $i<sizeof($a); $i++)
  echo $a[$i] . '<br />';
echo "Using implied keys with foreach... loop<br />";
foreach ($a as $i => $x)
  echo 'a[' . $i . '] = ' . $x . '<br />';

echo "An array with keys starting at an integer other than
0<br />";
$negKey = array(-1 => 'BMW', 'Lexus', 'house');
for ($i=-1; $i<2; $i++)
  echo $negKey[$i] . '<br />';

echo 'A keyed array with consecutive character keys...<br
/>';
$stuff = array('a' => 'BMW', 'b' => 'Lexus', 'c' =>
'house');
for ($i='a'; $i<='c'; $i++)
  echo $stuff[$i] . '<br />';
?>
```

```
Using for... loop
david
apple
Xena
Sue
Using implied keys with foreach... loop
a[0] = david
a[1] = apple
a[2] = Xena
a[3] = Sue
An array with keys starting at an integer other than 0
[–1] = BMW
[0] = Lexus
[1] = house
A keyed array with consecutive character keys...
[a] = BMW
```

> [b] = Lexus
> [c] = house

Document 3.2 demonstrates that even if specific key definitions are omitted, they still exist., and are given default integer values starting at 0. It also shows that it is possible to define just the first key, and the other keys will be assigned consecutively.

The ability to specify just the starting key provides an easy way to start array indices at 1 rather than 0, as might be convenient for labeling columns and rows in a table:

```
$a = array(1 => 63.7, 77.5, 17, -3);
```

The first index has a value of 1 and the remaining unspecified indices are incremented by 1. Either a `foreach`... or a `for`... loop can be used to access the values, as shown in Document 3.3.

Document 3.3. (base_1Array.php)

```php
<?php
echo '<br />A keyed array with indices starting at 1...<br
/>';
$a = array(1 => 63.7, 77.5, 17, -3);

foreach($a as $key => $val) {
  echo 'a[' . $key . '] = '. $val . '<br />';
}
for ($i=1; $i<=sizeof($a); $i++)
  echo $a[$i] . '<br />';
?>
```

```
A keyed array with indices starting at 1...
a[1] = 63.7
a[2] = 77.5
a[3] = 17
a[4] = -3
63.7
77.5
17
-3
```

Two-dimensional arrays—think of them as row-and-column tables—can be formed from an array of arrays, as shown in Document 3.4.

Document 3.4. (two-D.php)

```php
<?php
echo '<br />A 2-D array<br />';
$a = array(
  0 => array(1,2,3,4),
  1 => array(5,6,7,8),
  2 => array(9,10,11,12),
  3 => array(13,14,15,16),
  4 => array(17,18,19,20)
);
$n_r=count($a); echo '# rows = ' . $n_r . '<br />';
$n_c=count($a[0]); echo '# columns = ' . $n_c . '<br />';
for ($r=0; $r<$n_r; $r++) {
  for ($c=0; $c<$n_c; $c++)
    echo $a[$r][$c] . ' ';
  echo '<br />';
}
?>
```

```
A 2-D array
# rows = 5
# columns = 4
1 2 3 4
5 6 7 8
9 10 11 12
13 14 15 16
17 18 19 20
```

Document 3.4 uses the count() function to determine the number of rows and columns in the array; this function is completely equivalent to and interchangeable with sizeof(). The number of elements in $a, the "rows," is returned by count($a). Each element in $a is another array containing the "columns," and count($a[0]) (or any other index) returns the number of elements in this array. The count() function counts only defined array elements, so in order for it to work as expected, every element in an array must be given a value. In Document 3.4, defining the first row as

```
0 => array(1,2,3)
```

will result in the number of columns being identified as 3 rather than 4 if you use count($a[0]).

Higher-dimension arrays can be defined by extending the above procedure.

3.2 Array Sorting

PHP supports several functions for sorting arrays, including a `sort()`
function similar to JavaScript's `sort()` method. Consider Document 3.5.

Document 3.5. (`sort1.php`)

```php
<?php
// Create and sort an array…
$a = array('david','apple','sue','xena');
echo 'Original array:<br />';
for ($i=0; $i<sizeof($a); $i++)
  echo $a[$i] . '<br />';
sort($a);
echo 'Sorted array:<br />';
for ($i=0; $i<sizeof($a); $i++)
  echo $a[$i] . '<br />';
?>
```

This code produces the expected results with the array as defined:

```
Original array:
david
apple
sue
xena
Sorted array:
apple
david
sue
xena
```

But it won't do what you probably want for this change to the array, in
which two names are capitalized:

```php
$a = array('david','apple','Xena','Sue');
```

Recall that JavaScript's `sort()` method also produced
unexpected results because of its default actions in deciding which values
were "less than" others. For example, "Sue" is considered to be less than
"sue" because the uppercase alphabet comes earlier in the **ASCII**

character sequence than the lowercase alphabet. The same problems exist with PHP's basic `sort()` function:

```
Original array:
david
apple
Xena
Sue
Sorted array:
Sue
Xena
apple
david
```

JavaScript programmers will also be concerned about potential problems with arrays of numbers because of how JavaScript treats values passed from form fields. For example, because "1" comes before "3" in the ASCII character sequence, JavaScript's `sort()` will consider "13" to be less than "3" unless the `parseFloat()` or `parseInt()` methods are first applied to the values to convert them explicitly to numbers. Document 3.6 demonstrates that PHPs `sort()` function does not have this problem with numbers.

Document 3.6. (`sort2.php`)

```php
<?php
        $a=array(3.3,-13,-0.7,14.4);
        sort($a);
        for ($i=0; $i<sizeof($a); $i++)
          echo $a[$i] . '<br />';
?>
```

```
-13
-0.7
3.3
14.4
```

PHP offers several ways to sort arrays of strings and other combinations of elements, but a simple and reliable approach is to use the `usort()` function and provide a user-defined function that compares one array element against another using user-supplied criteria. This is comparable to using the JavaScript `sort()` function with its optional argument giving the name of a user-defined function to compare array elements. The user-supplied comparison function must return an integer value less than 0 if the first argument is to be considered as less than the second, 0 if they are equal, and greater than 0 if the first argument is greater than the second. For an array with strings containing upper- and lowercase letters, the very simple function shown in Document 3.7 makes use of strcasecmp() to perform a case-insensitive comparison of two strings.

Document 3.7. (sort3.php)

```php
<?php
function compare($x,$y) {

   return strcasecmp($x,$y);
}
// Create and sort an array...
$a = array('Xena', 'Sue', 'david', 'apple');
echo 'Original array:<br />';
for ($i=0; $i<sizeof($a); $i++)
   echo $a[$i] . '<br />';
echo 'Sorted array with user-defined comparisons of
elements:<br />';
usort($a,"compare");
for ($i=0; $i<sizeof($a); $i++)
   echo $a[$i] . '<br />';
?>
```

```
Original array:
Xena
Sue
david
apple
Sorted array with user-defined comparisons
of elements:
apple
david
Sue
Xena
```

3.3 Stacks, Queues, and Line Crashers

Stacks and queues are abstract data structures familiar to computer science students. They are used to store and retrieve data in a particular way. In PHP and other languages, stacks and queues can be represented as arrays. It is very easy to work with queues and stacks in PHP because, as with JavaScript, arrays can be resized dynamically, while a script is running.

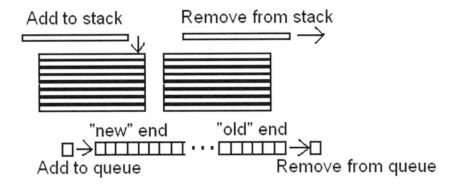

A stack uses a last-in/first-out (LIFO) data storage model. You can think of it as a stack of dinner plates. You put new dinner plates on the top of the stack, and when you retrieve a dinner plate, it always comes from the top of the stack. The last value added on a stack is the first value retrieved.

A queue uses a first-in/first-out (FIFO) data storage model. It operates like a queue of people waiting. A new person joins the line at the end, and people leave the line according to who has been in line the longest. A value removed from the queue is always the "oldest" value.

Because PHP's array model allows (assumes by default, actually) user-defined indices (keys) for each element, the possibilities for adding elements to and removing them from stacks or queues (or other array operations models) are complicated. This section will deal only with the basics, assuming arrays with integer indices that start at 0. This limited approach is sufficient for many science and engineering problems and for basic data handling.

The basic functions are `array_pop()`, `array_push()`, `array_shift()`, and `array_unshift()`. Document 3.8 demonstrates the use of each of these functions.

Document 3.8. (`StacksAndQueues.php`)

```
<html>
<head>
<title>Stacks and Queues</title>
</head>
```

```php
<body>
<?php
  $a = array(-17,"David", 33.3,"Laura");
  // Treat $a like a stack (last in, first out)…
  echo "The original array (element [0] is the \"oldest\"
  element):<br />";
  print_r($a);
  // Add two elements to $a…
  array_push($a,"Susan",0.5);
  echo "<br />Push two elements on top of stack:<br />";
  print_r($a);
  // Remove three elements from $a…
  array_pop($a); array_pop($a); array_pop($a);
  echo "<br />Remove three elements from top of stack:<br
  />";
  print_r($a);
  // Treat $a like a queue (first in, first out)…
  $a = array(-17,"David", 33.3,"Laura");
  echo "<br />Back to original array:<br />";
  print_r($a);
  echo "<br />Remove two elements from front of queue:<br
  />";
  array_shift($a);
  array_shift($a);
  print_r($a);
  echo "<br />Add three elements to end of queue:<br />";
  array_push($a,"Susan",0.5,"new_guy");
  print_r($a);
  echo "<br />Add a \"line crasher\" to the beginning of the
  queue:<br />";
  array_unshift($a,"queue_crasher_guy");
  print_r($a);
?>
</body>
</html>
```

```
The original array (element [0] is the "oldest"
element):
Array ( [0] => -17 [1] => David [2] => 33.3
[3] => Laura )
Push two elements on top of stack:
Array ( [0] => -17 [1] => David [2] => 33.3
[3] => Laura [4] => Susan [5] => 0.5 )
Remove three elements from top of stack:
Array ( [0] => -17 [1] => David [2] => 33.3 )
Back to original array:
Array ( [0] => -17 [1] => David [2] => 33.3
[3] => Laura )
Remove two elements from front of queue:
Array ( [0] => 33.3 [1] => Laura )
```

```
Add three elements to end of queue:
Array ( [0] => 33.3 [1] => Laura [2] => Susan
[3] => 0.5 [4] => new_guy)
Add a "line crasher" to the beginning of the queue:
Array ( [0] => queue_crasher_guy [1] => 33.3
[2] => Laura [3] => Susan
[4] => 0.5 [5] => new_guy )
```

Document 3.8 deserves close study if you need to do this kind of data manipulation in an array.

3.4 More Examples

3.4.1 The Quadratic Formula Revisited

In Document 1.6b (quadrat.php), three coefficients of a quadratic equation were passed from an HTML document and retrieved by name:

```
$a = $_POST["a"];
$b = $_POST["b"];
$c = $_POST["c"];
```

This code requires the PHP application to "know" what names the form fields were given in the corresponding HTML document ("a", "b", and "c"). In PHP terminology, you can think of the form fields being passed as a keyed array, with the key names corresponding to the form field names. For this and similar kinds of problems, it might be desirable to make the code less dependent on names given in the HTML document. Documents 3.9a and b show one way to do this.

Document 3.9a (quadrat2.htm)

```
<html>
<head>
<title>Solving the Quadratic Equation</title>
</head>
<body>
<form method="post" action="quadrat_2.php">
Enter coefficients for ax<sup>2</sup> + bx + c = 0:
<br />
a = <input type="text" value="1" name="coeff[0]" />
  (must not be 0)<br />
b = <input type="text" value="2" name="coeff[1]" /><br />
c = <input type="text" value="-8" name="coeff[2]" /><br />
<br /><input type="submit" value="click to get roots..." />
</form>
```

```
</body>
</html>
```

Document 3.9b (quadrat2.php)

```php
<?php
var_dump($_POST["coeff"]);
echo "<br />";
$coefficientArray=array_keys($_POST["coeff"]);
$a = $_POST["coeff"][$coefficientArray[0]];
$b = $_POST["coeff"][$coefficientArray[1]];
$c = $_POST["coeff"][$coefficientArray[2]];
$d = $b*$b - 4.*$a*$c;
if ($d == 0) {
   $r1 = $b/(2.*$a);
   $r2 = "undefined";
}
else if ($d < 0) {
   $r1 = "undefined";
   $r2 = "undefined";
}
else {
$r1 = (-$b + sqrt($b*$b - 4.*$a*$c))/2./$a;;
$r2 = (-$b - sqrt($b*$b - 4.*$a*$c))/2./$a;;
}
echo "r1 = " . $r1 . ", r2 = " . $r2;
?>
```

array(3) { [0]=> string(1) "1" [1]=> string(1) "2" [2]=> string(2) "-8" }
r1 = 2, r2 = -4

Document 3.9a is similar to Document 1.6a, but there is an important difference, marked with shaded code. Instead of each coefficient having a unique name, each one is assigned to an element of an array named coeff.

The calculations of the real roots in Document 3.9b are identical to those in Document 1.6b, but this code assumes that the PHP script does not automatically "know" the names of the quadratic coefficients, and that an array containing those coefficients may use keys other than consecutive integers starting at 0. Instead, the array_keys() function is used to extract the key names through the coeff[] array, available in $_POST, First, the values are placed in $coefficientArray, which uses default integer keys starting at 0. These values are then used as indices to the coeff array passed to the $_POST[] array.

If integer array keys starting at 0 are used, then the code for retrieving the coefficients can be simplified a little, as shown in the shaded lines of Documents 3.10a and b, which are otherwise identical to Documents 3.9a and b.

Document 3.10a (`quadrat3.htm`)

```
<html>
<head>
<title>Solving the Quadratic Equation</title>
</head>
<body>
<form method="post" action="quadrat3.php">
Enter coefficients for ax<sup>2</sup> + bx + c = 0:
<br />
a = <input type="text" value="1" name="coeff[]" />
  (must not be 0)<br />
b = <input type="text" value="2" name="coeff[]" /><br />
c = <input type="text" value="-8" name="coeff[]" /><br />
<br /><input type="submit" value="click to get roots..." />
</form>
</body>
</html>
```

Document 3.10b (quadrat3.php)

```
<?php
var_dump($_POST["coeff"]);
echo "<br />";
$coefficientArray=$_POST["coeff"];
$a = $coefficientArray[0];
$b = $coefficientArray[1];
$c = $coefficientArray[2];
$d = $b*$b - 4.*$a*$c;
if ($d == 0) {
  $r1 = $b/(2.*$a);
  $r2 = "undefined";
}
else if ($d < 0) {
  $r1 = "undefined";
  $r2 = "undefined";
}
else {
$r1 = (-$b + sqrt($b*$b - 4.*$a*$c))/2./$a;;
$r2 = (-$b - sqrt($b*$b - 4.*$a*$c))/2./$a;;
}
echo "r1 = " . $r1 . ", r2 = " . $r2;
?>
```

Note that because no index values are specified for the `coeff[]` array in Document 3.10a, PHP assumes that they are integers starting at 0.

You could also specify the keys, for example, as integers starting at 1, but you must then tell the PHP application what the keys are, either by using `array_keys()` or by "hard coding" the key values:

(in the HTML document)
```
a = <input type="text" value="1" name="coeff[1]" />
  (must not be 0)<br />
b = <input type="text" value="2" name="coeff[2]" /><br />
c = <input type="text" value="-8" name="coeff[3]" /><br />
```

(in the PHP document)
```
$coefficientArray=$_POST["coeff"];
$a = $coefficientArray[1];
$b = $coefficientArray[2];
$c = $coefficientArray[3];
```

This code is easier to write with consecutive integer keys than with arbitrarily named keys, but this approach essentially defeats the purpose of simplifying access to form fields, which the example was originally intended to demonstrate.

You might conclude that the code presented in these examples is not much of a simplification and is not worth the extra effort, because the PHP document still needs to "know" the name of the coefficient array entered into the HTML document's form fields. However, if only one name is needed—the name of that array—this might provide some code-writing economy for a longer list of inputs.

3.4.2 *Reading* checkbox *Values*

The HTML `<input type="checkbox" … />` form field is used to associate several possible values with a single form field name. Unlike a `type="radio"` field, which allows only one selection from a list, checkboxes allow multiple values to be selected. Consider this problem:

An HTML document asks a user to report cloud observations by checking boxes for cloud types divided into four categories: high, mid, low, and precipitating. Each category has more than one possible cloud type, and multiple cloud types in one or all categories may be observed:

High: cirrus, cirrocumulus, cirrostratus
Mid: altostratus, altocumulus
Low: stratus, stratocumulus, cumulus

Precipitating: nimbostratus, cumulonimbus

Write an HTML document to enter these data and a PHP document that will echo back all the cloud types reported.

Document 3.11a (CloudObs.htm)

```html
<html>
<head>
<title>Cloud Observations</title>
</head>
<body bgcolor="#aaddff">
<h1>Cloud Observations</h1>
<strong> Cloud Observations </strong>(Select as many cloud
types as observed.)
<br />
<form method="post" action="CloudObs.php" />
<table>
  <tr>
   <td><strong>High</strong> </td>
    <td>
     <input type="checkbox" name="high[]"
       value="Cirrus" /> Cirrus</td>
    <td>
     <input type="checkbox" name="high[]"
       value="Cirrocumulus" /> Cirrocumulus </td>
    <td>
      <input type="checkbox" name="high[]"
       value="Cirrostratus" /> Cirrostratus </td></tr>
  <tr>
    <td colspan="4"><hr noshade color="black" />
     </td></tr>
  <tr>
    <td> <strong>Middle</strong> </td>
    <td>
      <input type="checkbox" name="mid[]"
       value="Altostratus" /> Altostratus </td>
    <td>
      <input type="checkbox" name="mid[]"
       value="Altocumulus" /> Altocumulus</td></tr>
  <tr>
    <td colspan="4"><hr noshade color="black" />
     </td></tr>
  <tr>
    <td> <strong>Low</strong></td>
    <td>
    <input type="checkbox" name="low[]" value="Stratus" />
      Stratus</td>
    <td>
      <input type="checkbox" name="low[]"
```

```
        value="Stratocumulus" /> Stratocumulus</td>
    <td>
    <input type="checkbox" name="low[]" value="Cumulus" />
        Cumulus </td></tr>
  <tr>
    <td colspan="4"><hr noshade color="black" />
        </td></tr>
  <tr>
    <td> <strong>Rain-Producing </strong> </td>
    <td>
        <input type="checkbox" name="rain[]"
            value="Nimbostratus" /> Nimbostratus</td>
    <td>
        <input type="checkbox" name="rain[]"
            value="Cumulonimbus" /> Cumulonimbus </td></tr>
</table>
<input type="submit" value="Click to process..." />
</form>
</body>
</html>
```

Cloud Observations

Cloud Observations (Select as many cloud types as observed.)

| High | ☐ Cirrus | ☑ Cirrocumulus | ☑ Cirrostratus |

| Middle | ☐ Altostratus | ☐ Altocumulus |

| Low | ☐ Stratus | ☐ Stratocumulus | ☐ Cumulus |

Rain-Producing ☐ Nimbostratus ☑ Cumulonimbus

Click to process...

It is very easy to process these data with PHP if the HTML document is written correctly. Each cloud category—high, mid, low, or precipitating—must be specified as an array high[] rather than just high, for example. (Note that you do not need to specify the index values.) The $_POST[] operation performed in PHP will return an array including just those cloud types that have been checked. PHP automatically does the work that would require you to write more code in JavaScript. The PHP code to do this is given in Document 3.11b, below.

Document 3.11b (`CloudObs.php`)

```php
<?php
   $high = $_POST["high"];
   $n = count($high);
   echo "For high clouds, you observed<br />";
   for ($i=0; $i<$n; $i++)
      echo $high[$i] . "<br>";
   $mid = $_POST["mid"];
   $n = count($mid);
   echo "For mid clouds, you observed<br />";
   for ($i=0; $i<$n; $i++)
      echo $mid[$i] . "<br />";
   $low = $_POST["low"];
   $n = count($low);
   echo "For low clouds, you observed<br />";
   for ($i=0; $i<$n; $i++)
      echo $low[$i] . "<br />";
   $rain = $_POST["rain"];
   $n = count($rain);
   echo "For precipitating clouds, you observed<br />";
   for ($i=0; $i<$n; $i++)
      echo $rain[$i] . "<br />";
?>
```

> For high clouds, you observed
> Cirrocumulus
> Cirrostratus
> For mid clouds, you observed
> For low clouds, you observed
> For precipitating clouds, you observed
> Cumulonimbus

The number of boxes checked for each category is contained in the value of $n, which is reset after each $_POST[]. For mid and low clouds, no boxes are checked, so their corresponding arrays are empty and their for... loops are not executed. It would be a simple matter to use the value of $n to determine whether the message displayed for an empty category should be different; for example, "There were no low clouds observed."

3.4.3 Building a Histogram Array

> Write a PHP application that reads scores between 0 and 100 (possibly including both 0 and 100) and creates a histogram array whose elements contain the number of scores between 0 and 9, 10 and 19, etc. The last "box" in the histogram should include scores between 90 and 100. Use a function to generate the histogram.

The solution shown here is a minimal approach to this problem. It assumes that the range of the values is from 0 to some specified number, and that the histogram "boxes" are uniformly distributed over the range. The data file looks like this:

```
73
77
86
...
17
18
```

Your application should not assume that the number of entries in the file is known ahead of time.

Document 3.12 (histo.php)

```php
<?php
  function buildHisto($a,$lo,$hi,$n_boxes) {
    echo "building histogram...<br />";
    $h=array();
    // echo "Number of boxes = ".$n_boxes."<br />";
    for ($i=0; $i<$n_boxes; $i++) {
        array_push($h,0);
        // echo $h[$i]."<br />";
    }
      echo "size of histogram array = ".sizeof($h)."<br />";
      for ($n=0; $n<sizeof($a); $n++) {
        $i=floor($a[$n]/$n_boxes);
        if ($i==sizeof($h)) $i--;
        $h[$i]++;
      }
      $sum=0;
      for ($i=0; $i<sizeof($h); $i++) {
        echo "h[".$i."] = ".$h[$i]."<br />";
        $sum+=$h[$i];
      }
      echo "# of entries = ".$sum."<br />";
  }
```

```
$in=fopen("histo.dat","r");
$a=array();
$i=0;
while (!feof($in)) {
   fscanf($in, "%f", $s);
      $a[$i]=$s;
      $i++;
      // array_push($a,$s); will also work.
      // echo 'a['.$i.'] = '.$a[$i].'<br />';
   }
/* Alternative code…
   $i=0;
   while (!feof($in)) {
      fscanf($in,"%f",$a[$i]);
      $i++;
   }
*/
   echo 'Number of scores: '.sizeof($a).'<br />';
   buildHisto($a,0,100,10);
   fclose($in);
?>
```

```
Number of scores: 39
building histogram...
size of histogram array = 10
h[0] = 1
h[1] = 5
h[2] = 2
h[3] = 5
h[4] = 5
h[5] = 1
h[6] = 3
h[7] = 3
h[8] = 8
h[9] = 6
# of entries = 39
```

The first step is to open and read the data file. This code demonstrates how to store values in an array as they are read, one at a time, from a data file. Each value is read into a variable and that variable is then assigned to the appropriate array element. This is a good approach if there is a reason to test or modify the value read before saving it in the array. Alternative code is also shown which reads every value directly into an array element. An `echo` statement included to display the values as they are read, during code testing, is later commented out of the script.

Next, the histogram function is called. In addition to the number of histogram bins, the lower and upper limits to the range of values are provided in case the code needs to be modified later to accommodate data values that don't have 0 as their lower limit. In this simple solution, the lower and upper limits on the range of the values are not needed.

In function `buildHisto()`, the contents of each "bin" are initialized to 0. The array index value for the histogram array is calculated as `$i=floor($a[$n]/$n_boxes);`. In the case of a score of 100, this index would have a value of 10, which is beyond the 10 allowed boxes (indices 0–9), so in this case the index value is reduced by 1 and that value is put in the box holding values from 90 to 100. This simple calculation of histogram array indices is possible only because the original data values lie between 0 and 100. In general, a more sophisticated calculation would be required to associate values with the appropriate histogram array element.

The `buildHisto()` function includes some `echo` statements for testing which are later commented out. It is very important to include these intermediate outputs whenever you are developing a new application, to ensure that your code is doing what you expect it to do.

3.4.4 Shuffle a Card Deck

> Write a PHP application that will shuffle a deck of 52 "cards." Represent the cards as an array of integers having values from 1 to 52. After the results of shuffling this "deck" are displayed, sort the deck in ascending order and display it again.

This is a very simple statement of a random shuffling problem. The solution presented is to read once through the deck and exchange each element with another randomly chosen element in the array, using the `rand()` function to select the element. Note that you cannot simply use the `rand()` function to generate 52 random "cards," because, almost always, some card values will be duplicated and some will not appear at all.

Document 3.13 (`cardShuffle.php`)

```php
<?php
  $deck = array();
  for ($i= 0; $i<52; $i++) {
```

```
    $deck[$i]=$i+1;
        echo $deck[$i]." ";
}
echo "<br />";
for ($i=0; $i<52; $i++) {
    $j=rand(0,51);
        $save=$deck[$i];
        $deck[$i]=$deck[$j];
        $deck[$j]=$save;
}
for ($i=0; $i<52; $i++)
    echo $deck[$i]." ";
echo "<br />";
sort($deck);
echo "Resort deck...<br />";
for ($i=0; $i<52; $i++)
    echo $deck[$i]." ";
echo "<br /";
?>
```

> 1 2 3 4 5 6 7 8 9 10 11 12 13 14 15 16 17 18 19 20 21 22 23 24
> 25 26 27 28 29 30 31 32 33 34 35 36 37 38 39 40 41 42 43 44 45
> 46 47 48 49 50 51 52
> 17 6 23 38 22 28 49 40 10 11 33 36 5 25 4 31 30 7 2 15 47 12 46
> 29 16 26 8 37 44 19 41 45 35 34 52 1 43 13 21 39 27 48 24 14
> 50 32 20 42 18 3 9 51
> Resort deck...
> 1 2 3 4 5 6 7 8 9 10 11 12 13 14 15 16 17 18 19 20 21 22 23 24
> 25 26 27 28 29 30 31 32 33 34 35 36 37 38 39 40 41 42 43 44 45
> 46 47 48 49 50 51 52

The multiple echo statements in the code show the results of the code at each step. The purpose of resorting the deck and displaying the results is to make sure that the code actually moves the original elements around and does not, for example, overwrite elements in a way that might produce duplicate or missing values.

3.4.5 Manage a Data File

> Write an HTML/PHP application that allows you to manage entries in a text file stored on a server. The application should be able to:
>
> (a) display all records;

(b) look for a specified date or value in the file;
(c) insert a new data report into the file;
(d) optionally, remove a record;
(e) optionally, reset the file to a previous version.

There can be duplicate values and/or dates, so you must look for all of them. When you insert a new data report into the file, it must be inserted into its chronologically correct position in the file.

To determine where a new record goes, or to look for a requested date, you need to know whether the date of a record is later than ("greater than"), the same as ("equal to"), or earlier than ("less than") some other record. You can do this with the strtotime($date) function, which converts a date given in any reasonable format, including the mm/dd/yyyy format shown in the sample file below, into the integer number of seconds since (probably) 01 January, 1970.[1] Hence, this function allows you to compare dates as needed. For example, to see if you have matched a requested date $date to a date $d in the file,

```
if (strtotime($d) == strtotime($date)) {...
```

The HTML document interface for this program should look something like this:

Date (mm/dd/yyyy format): 01/01/2007

Value (number): 17.7

Find date: ⌒   Find value: ⌒

Insert new report (in chronological order): ⌒   View all reports: ⊙

Submit request

Your PHP application must process the radio button selection. This is easy, because $_POST["{name of your radio button}"] returns the text value of the value attribute of the selected button.

Here is a sample initial file. It consists of a header line, followed by a series of data entries consisting of a date in mm/dd/yyyy format and a value separated by a space.

---

[1] This is the usual reference date, but it is certainly possible for some operating systems to choose a different reference. In any event, the starting date does not matter.

Date Value
01/15/2006 17.3
01/20/2006 0.55
05/17/2006 83.9
09/09/2006 9.33
11/13/2006 15
01/01/2007 74.4
02/28/2007 64.4
05/05/2007 100
06/06/2007 64.4
12/12/2007 22.54

The "insert new data" option for this program will require you to open a file for reading, close it, and then open it again for writing in order to insert the new record in its appropriate place. Because these are sequential access files—opened only for reading or writing, but not both at the same time—you cannot simply write the new record into your existing open file. Instead, you need to read the data in the file into an array, insert the new record into its appropriate position in the array, close the original file, and then open the file again in "write" mode so you can copy the expanded array back into the original file. Alternatively, you might write all records, including a new record, directly into a temporary file and then copy this new file back to the original file.

```
if (!copy("c:/Documents and Settings/All Users/
    Documents/phpout/values.out",
    "c:/Documents and Settings/All Users/
    Documents/phpout/values.dat") )
echo "Failed to copy file.";
```

This solution would not require the use of arrays at all. In fact, depending on what you want to do with the data stored in your file, you might not need arrays for any of the options specified for this application.

In any case, you will need to establish appropriate access permissions for the folder in which files will be written. On Windows computers, these permissions are set in the "file sharing" options for a folder. The default permissions may allow you to read files and create new files, but not to replace an existing file. If the permissions are not set correctly, you may get a message something like, "This resource stream is not available..." Such a message relates not necessarily to your code, but to how your computer operating system manages file access.

Documents 3.14a and b show how to set up a basic data file management application.

Document 3.14a (DataReport.htm)

```html
<html>
<head>
<title>Manage Data File</title>
</head>
<body>
<h3>File Management Application</h3>
<form method="post" action="DataReport.php">
        Date (mm/dd/yyyy format): <input type="text"
name="date" value="01/01/2007" /><br />
        Value (number): <input type="text" name="value"
value="17.7" /><br />
        Find date: <input type="radio" name="act"
value="find_date" />
              Find value: <input type="radio"
name="act" value="find_value" /><br />
        Insert new report (in chronological order): <input
type="radio" name="act" value="insert_new" />
              View all reports: <input
type="radio" name="act" checked value="view_all" /><br />
        <input type="submit" value="Submit request" />
</form>
</body>
</html>
```

Document 3.14b (DataReport.php)

```php
<?php
  echo "In DataReport.php<br />";
  $date=$_POST["date"];
  $value=$_POST["value"];
  $selected=$_POST["act"];
  echo "Values passed: ".$date.", ".$value.",
        ".$selected,"<br />";
  $FileName="c:/Documents and Settings/All Users/
            Documents/phpout/values.dat";
  switch ($selected) {
// View all the records.
    case "view_all";
      echo "View all entries.<br />";
      $in=fopen($FileName,"r");
      while (!feof($in)) {
        if (!feof($in)) echo fgets($in)."<br />";
      }
      fclose($in);
      break;
```

```php
// Insert a new record.
    case "insert_new":
        $a=array();
        $in=fopen($FileName,"r");
        $header=fgets($in);
        $a[0]=$header;
        echo "header ".$a[0]."<br />";
        $newline=$date." ".$value;
        echo "Insert this line: ".$newline."<br />";
        $i=0;
        $found=false;
        while (!feof($in)) {
          fscanf($in,"%s %f",$d,$v);
          if (!feof($in)) {
            echo $d.", ".$v.", ".strtotime($d)."<br />";
            $line=$d." ".$v;
            $i++;
            if((strtotime($date)<=strtotime($d))&&(!$found)) {
              echo "new value goes before this one<br />";
              $a[$i]=$newline; $i++; $found=true;
            }
          $a[$i]=$line;
          }
        }
        echo "return from fclose(): ".fclose($in)."<br />";
        $out=fopen("c:/Documents and Settings/
           All Users/Documents/phpout/values.dat","w")
          or exit("Can't open file.");
        fprintf($out,"%s",$a[0]);
        for ($i=1; $i<sizeof($a); $i++) {
          echo "a[".$i."] = ".$a[$i]."<br />";
          fprintf($out,"%s\r\n",$a[$i]);
        }
        fclose($out);
        break;
// Find a specified value in the file.
    case "find_value":
        $in=fopen($FileName,"r");
        $found=false;
        echo "Looking for ".$value."...<br />";
        while (!feof($in)) {
          fscanf($in,"%s %f",$d,$v);
          if ($value==$v) {
            echo $d.", ".$v."<br />";
            $found=true;
          }
        }
        if (!$found) echo "Could not find this value.<br />";
        fclose($in);
        break;
// Find a specified date in the file.
    case "find_date":
```

```
$in=fopen($FileName, "r");
$found=false;
echo "Looking for date ".$date."...<br />";
while (!feof($in)) {
  fscanf($in, "%s %f",$d,$v);
  if (strtotime($d)==strtotime($date)) {
    echo $d.", ".$v."<br />";
    $found=true;
  }
}
if (!$found) echo "Could not find this date.<br />";
  fclose($in);
  break;
default: echo "No match.<br />";
}
?>
```

In this code, the file is opened at the beginning of each switch() case and closed at the end of processing for that case. For searching for a value or date, or for reviewing all entries in the file, this might not seem to be necessary, as all these operations are "read only" on the file. However, this is done so the PHP application will never leave any open files when you return to the HTML document to select a different option. Conceptually, at least, each "call" to the PHP application should be considered as restarting that application from the beginning.

An option to remove a record is not included in Document 3.14, although it is easy to do. Also not included is an option to reset a modified file back to some earlier version, although that is easily done by making a copy of the existing data file before adding (or removing) a record. You can always "undo" a change by copying that file over the modified file. Just be careful that you don't undo something you wished to save!

In principle, this application could be written with random rather than sequential access file structures (which might make inserting or removing records easier), or by using PHP-accessible databases, but those discussions are beyond the scope of this book. Although the simple approach implemented in Document 3.14 might be unwieldy for very large and complicated data files, it is perfectly satisfactory for small and simple files that you might need to maintain for your own work for nothing more complex than basic recordkeeping and table lookups.

4 Summary of Selected PHP Language Elements

This chapter provides a summary of PHP language elements that are necessary or helpful to create the kinds of applications that have been discussed in this book. As with other parts of this book, the descriptions are neither comprehensive nor necessarily complete, and they are not intended to take the place of a reference manual. Nonetheless, the language elements and examples presented in this chapter cover a large range of practical PHP programming situations.

4.1 Data Types and Operators

4.1.1 Data Types

PHP supports four scalar primitive data types:

boolean *(bool)*
integer *(int)*
float *(float)*
string *(string)*

Boolean data can have values of `true` or `false`. The maximum size of an integer that can be represented is system dependent, but integers are often represented with a 32-bit word, with one bit allocated for a sign. This gives a maximum integer value of 2,147,483,647. If presented with an integer larger than the allowed maximum, PHP will convert it to a floating point number, possibly with some loss of precision. The precision of floating-point numbers is also system dependent but is often approximately 14 digits. Occasionally you will find references to a "double" data type. In C, for example, the precision of "float" and "double" real-number values is different, but there is no such distinction in PHP, which supports only a single floating-point number representation. Strings are composed of 8-bit characters (giving 256 possible characters), with no imposed limit on string length.

D.R. Brooks, *Introduction to PHP for Scientists and Engineers*,
doi: 10.1007/978-1-84800-237-1_4, © Springer-Verlag London Limited 2008

PHP supports arrays and objects as compound data types. This book deals only with the *(array)* type, which can be used to aggregate a mixture of data types under a single name. Array elements can be any of the primitive data types, as well as other arrays.

When a collection of data with various types is specified, such as the elements of a mixed-type array, they can be identified for convenience as *(mixed)*, but this word represents only a **pseudo data type**, not an actual data type specification. In the definitions of math functions given later (see able 4.5), inputs and outputs are sometimes identified as having a *(number)* data type. This is a pseudo data type that can be either an integer or a floating-point number, depending on context.

Another pseudo data type is *(resource)*. This refers to any external **resource**, usually a data file, which is accessible by a PHP application.

4.1.2 Operators

PHP supports a large number of operators, some of which are listed in Table 4.1, in order of precedence.

Table 4.1 Operators, in decreasing order of precedence

Operator	Description
++, --	Increment/decrement
*, /, %	Multiplication, division, modulus division
+, -, .	Addition, subtraction, string concatenation
<, <=, >, >=	Relational comparisons
==, !=, ===, !==	Relational comparisons
&&[1]	Logical AND
||[1]	Logical OR
=, +=, -=, *=, /=, %=, .=	Arithmetic assignment, string concatenation assignment
and	Logical AND
xor	Logical EXCLUSIVE OR
or	Logical OR

[1] Note the availability of two AND (&& and and) and two OR (\|\| and or) operators, at different precedence levels. (and and or have lower precedence than && and \|\|.)

As always, it is good programming practice, especially in relational and logical expressions, to use parentheses to clarify the order in which operations should be performed, rather than depending solely on precedence rules.

4.2 Conditional Execution

PHP supports if... then... else... and case-controlled conditional execution. The "then" action is implied. Multiple "else" branches can be included. The example below illustrates typical syntax.

4.2.1 if... then... else... *Conditional Execution*

Document 4.1 (`conditionalExecution.php`)

```php
<?php
  function getRoots($a,$b,$c) {
    echo "This function calculates roots...";
  }
  $i = 2;
  if ($i == 0) {
    echo "i equals 0";
  }
  elseif ($i == 1) {
    echo "i equals 1";
  }
  elseif ($i == 2) {
    echo "i equals 2";
  }
  else {
    echo "i is not 0, 1, or 2";
  }
  echo "<br />";
  $discriminant=0.3;
  if ($discriminant < 0.)
    echo "There are no real roots.<br />";
  elseif ($discriminant == 0.) {
    echo "There is one real root.<br />";
    $r1 = -$b/$a/2;
    echo $r1;
  }
  else {
    echo "There are two real roots.<br />";
    list($r1,$r2) = getRoots($a,$b,$c);
    echo "<br />Print the roots here...";
  }
?>
```

> i equals 2
> There are two real roots.
> This function calculates roots...
> Print the roots here...

4.2.2 Case-Controlled Conditional Execution

PHP also has a "switch" construct for conditional execution.

```php
switch ($i) {
  case 0:
    echo "i equals 0.";
    break;
  case 1:
    echo "i equals 1.";
    break;
  case 2:
    echo "i equals 2.";
    break;
  default:
    echo "i does not equal 0, 1, or 2.";
}
```

The order of the case values does not matter. Unlike the if...
construct, in which only the first "true" path is executed, the break;
statement is needed to exit the construct after the first case match is
encountered. Otherwise, all subsequent statements within the construct are
executed. There are certainly circumstances under which this might be the
desired result, in which case the break; statements wouldn't be needed,
although the order of the case values probably *would* matter.

Multiple case values can be associated with the same action, as
shown in Document 4.2.

Document 4.2 (daysInMonth.php)

```php
<?php
$month=5;
switch ($month) {
  case 1:
  case 3:
  case 5:
  case 7:
  case 8:
  case 10:
```

```
  case 12:
    echo "There are 31 days in this month.<br />"; break;
  case 4:
  case 6:
  case 9:
  case 11:
    echo "There are 30 days in this month.<br />"; break;
  case 2:
    echo "There are either 28 or 29 days in this month.
<br />"; break;
  default:
    echo "I do not understand your month entry.";
}
?>
```

In PHP, case values can be strings:

```
switch ($fruit)
case "apple":
    echo "This is an apple.";
    break;
  case "orange":
    echo "This is an orange.";
    break;
  case "banana":
    echo "This is a banana.";
    break;
  default:
    echo "This is not an allowed fruit treat.";
}
```

Comparisons against the value to be tested are case-sensitive. So, if $fruit is assigned as $fruit = "Banana"; prior to the switch construct, (instead of $fruit = "banana";) the default message is printed. If this is a problem, it can be overcome by using the strtolower() or strtoupper() functions.

4.3 Loops

PHP supports both count-controlled and conditional execution loops, including a foreach... loop designed specifically for accessing keyed arrays. In the examples below, generic variable names such as *$counter* are sometimes used, displayed in *italicized Courier font*.

Programmer-supplied text and/or statements are represented by *{italicized Times Roman font in curly brackets}*.

4.3.1 Count-Controlled Loops

The basic count-controlled loop iterates over specified values. The general syntax is:

```
for ((int) $counter = $startValue;
    $counter {relational operator} $endValue;
    $counter = $counter {+ or –} $incrementValue) {
    {statements}
}
```

The statement(s) inside the loop are executed only if (or as long as) the second expression evaluates as true. As a result, it is possible that the statements(s) inside the loop may never be executed. The $startValue can be smaller or larger than the $endValue, as long as the third statement increments or decrements the $counter so that the loop will eventually terminate (that is, the second expression evaluates as false). With appropriately defined conditions, the loop can count "backwards." The curly brackets are optional if there is only one statement to be executed inside the loop.

Example:

Document 4.3 (countdown.php)

```
<?php
  for ($i=10; $i>=0; $i--)
    echo $i . "<br />";
  echo "FIRE!<br />";
?>
```

| 10 |
| 9 |
| 8 |
| 7 |
| 6 |
| 5 |
| 4 |
| 3 |
| 2 |
| 1 |
| FIRE! |

Document 4.4 (loopExamples.php)

```
<?php
  $a = array(17,-13.3, "stringThing","PHP");
  foreach ($a as $x)
    echo "$x<br />";
  for ($i=0; $i<=sizeof($a); $i++)
    echo $a[$i] . '<br />';
  $a = array(1 => 17,2 => -13.3, 3 => "stringThing",
```

```
    4 => "PHP");
  foreach ($a as $k => $x)
    echo "a[" . $k . "] = " . $x . "<br />";
  $b = array(77, 33, 4);
  foreach($b as $x) {
    echo("$x" . "<br />");
  }
?>
```

```
17
–13.3
stringThing
PHP
17
–13.3
stringThing
PHP

a[1] = 17
a[2] = –13.3
a[3] = stringThing
a[4] = PHP
77
33
4
```

See Document 3.2 for an example of how to use a for... loop to access an array with character indices. Document 3.2 also shows that it is possible to define just the first key value in an array, with the remaining keys automatically assigned with consecutive values.

The foreach... loop is used to access keyed elements in an array, including arrays with other than integer indices. The curly brackets are optional if there is only one statement.

```
foreach ( (array) $a as $value) {
  {one or more statements}
}

foreach ( (array) $a as $key => $value) {
  {one or more statements}
}
```

Example:

Document 4.5 (`foreach.php`)

```php
<?php
  $a = array(17,-13.3,
    "stringThing","PHP");
  foreach ($a as $x)
    echo "$x<br />";

  $a = array(1 => 17,2 => -13.3,
    3 => "stringThing", 4 => "PHP");
  foreach ($a as $k => $x)
    echo "a[" . $k . "] = " . $x . "<br />";
?>
```

```
17
-13.3
stringThing
PHP
a[1] = 17
a[2] = -13.3
a[3] = stringThing
a[4] = PHP
```

4.3.2 Condition-Controlled Loops

```
do (
    {one or more statements}
} while ( (bool) {logical expression});
```

The conditional do... loop executes statements as long as the *{logical expression}* evaluates as true. Because the expression is evaluated at the end of the loop, the statements inside the loop will always be executed at least once.

```
while ({logical expression}) {
    {one or more statements}
}
```

The conditional while... loop executes statements as long as the *{logical expression}* evaluates as true. Because the expression is evaluated at the beginning of the loop, it is possible that the statements inside a `while...` loop will never be executed.

Examples:

Document 4.6 (`squares.php`)

```php
<?php
$x=0;
do {
  $x++;
  echo $x . ', ' . $x*$x . '<br />';
} while ($x*$x < 100.);
?>
```

```
1, 1
2, 4
3, 9
4, 16
5, 25
6, 36
7, 49
8, 64
9, 81
10, 100
```

```php
<?php
  $in = fopen("stuff.dat";, "r") or
    exit("Can't open file stuff.dat.");
  while (!feof($in)) {
    $line=fgets($in);
    echo $line . "<br />";
  }
  fclose($in);
?>
```

4.4 Functions and Constructs

There are literally hundreds of PHP functions and language constructs. This section contains a subset of functions and constructs used in or closely related to those used in this book. In these descriptions, the data type of an input parameter or return value is given in italicized parentheses, *e.g.*, *(string)*. Programmer-supplied text is printed in *{italicized Times Roman font}* inside curly brackets. Often, generic variable names are given in *italicized Courier font*, e.g., *$fileHandle*. Optional parameters are enclosed in square brackets.

4.4.1 File Handling and I/O Functions

As noted previously, file access is the primary justification for using a server-side language such as PHP. As a general rule, you can read files from anywhere on a local computer, but you need appropriate access permissions to write or modify files. On Windows computers, you can set permissions for a folder, including giving permission for network users to change the contents of files, by right-clicking on the folder and selecting "Properties" and then "Sharing." Past that general advice, you may need to need to ask your system administrator about these matters. Problems usually manifest themselves when PHP refuses to open a file in write (`"w"`) or append (`"a"`) mode.

Format specifiers

Many of the PHP I/O functions described below which read input or display output require format specifiers that control how input is interpreted and how output is displayed. Each output format conversion specifier starts with a percent sign (`%`) followed by, in order, one or more of these *optional* elements:

Sign specifier

Either a "–" or a "+" forces numbers to be displayed with a leading sign. (By default, negative numbers are preceded by a "–" sign, but positive numbers are not preceded by a "+" sign.)

Padding specifier

The padding specifier is a character, preceded by a single quote ('), used for padding numerical results to the appropriate string size. The default character is a space. A typical non-default character would be a 0.

Alignment specifier

By default, output is right-justified. Including a "–" will force left justification.

Width specifier

A numerical width specifier defines the minimum number of spaces allocated for display of a number or string. If the width specifier is too small, it will be overridden to allow display of the entire number or string.

Precision specifier

A numerical precision specifier, preceded by a decimal point, defines how many digits to the right of the decimal point should be displayed for floating-point numbers. Often used along with the width specifier, for example, 8.3. When applied to a string, this value defines the maximum number of characters displayed. When significant digits are lost, the result is rounded rather than truncated. For example, an *n*.3 specifier applied to 17.4567 will display the number as 17.457. Numbers are right-padded with 0's. For example, an *n*.3 format specifier applied to 17.5 will display 17.500.

Type specifier

A required type specifier for input or output defines how an argument should be interpreted and displayed. Some type specifiers for strings and base-10 numbers are given in Table 4.2. Format strings can contain characters other than the type specifiers themselves. For example, the statement

```
fscanf($inFile,"%u,%f",$i,$x);
```

implies that a line in the input file contains an integer and a floating-point number, with a comma directly after the integer. The number of spaces between a comma and the following number does not matter. For example, it does not matter whether two values in a file are stored as

```
17,33.3
```

or

```
17,             33.3
```

The same format specifier used for output would display one integer and one floating-point number in default format, separated by a comma. With output, multiple spaces embedded in a format string are collapsed into a single space when they are displayed in a browser, but those spaces are retained if the output is sent to a file.

Table 4.2 Selected type specifiers for input and formatted output

Specifier	Description
c	Treats argument as an integer, displays it as the character having that base-10 ASCII value
d	Displays numerical value as a signed base-10 integer
e or E	Displays numerical value in scientific notation, for example, 7.444e-3
f	Reads a value as a floating-point number. Displays numerical value as a floating-point number
s	Reads a value as a string. Displays argument as a string
u	Reads a value as a base-10 integer. Displays numerical value as a base-10 integer

Table 4.3 gives some **escape sequences** that are preceded by a backslash, the **escape character**, or a percent sign within format strings.

Table 4.3 Selected escape characters

Escape character	Description
\n	Insert linefeed, ASCII base-10 value 10
\r	Insert carriage return, ASCII base-10 value 13
\t	Insert tab, ASCII base-10 value 9
\$	Display dollar sign
\'	Display single quote
\"	Display double quote
%%	Display percent character
\\	Display backslash character

```
$fileHandle = fopen((string) $fileName,
            (string) {mode})

(bool) fclose($fileHandle)
```

Associates a specified file name, its "physical name," with a specified access mode and assigns it to a user-supplied "file handle" represented by the generic resource name $fileHandle. The $fileHandle is used in the script as the file's "logical name." The $fileName can be a variable name assigned an appropriate value or a string literal. The {mode}, which specifies how the file can be accessed, is usually a string literal. The mode string can be surrounded by either single or double quotes.

fclose() closes a previously opened file pointed to by a file handle.

Example:

```
$inFile = "dataFile.dat";
$in = fopen($inFile, 'r');
$out = fopen("outputFile.dat", 'w');
```

There are several possible modes for opening files, including binary files, but this book assumes that all files are text files and will be opened either for reading or for writing, but not for both simultaneously. The three modes are summarized in Table 4.4.

Table 4.4 Text file access modes for fopen()

Mode	Description
"r"	Open for reading only, starting at beginning of file
"w"	Open for writing only. If file exists, its contents will be overwritten. If not, it will be created
"a"	Open for writing only, starting at the end of an existing file. If the file does not it exist, it will be created

For these access modes, all files are **sequential access** as opposed to **random access**. When a file is opened in read-only mode, its file handle points to the location of the first byte of the file in memory. Reading from such a file implies that you must read all the contents of the file, starting at the beginning, even if you discard some of the information. You cannot jump to a random location within the file.

In write-only mode, data are written to the file sequentially, starting at the beginning of a blank file. If the file handle represents a physical file that already exists, then the old file is replaced by the new data. (Be careful!) In append mode, the pointer to the file in memory is positioned initially at the end of the file, just before the end-of-file character. (In write-only or read-only mode, the file pointer is positioned initially at the beginning of the file.) New data are added to the end of the file without changing whatever was previously in the file.

(bool) feof((resource) $fileHandle)

Tests for the end-of-file marker on $fileHandle. Returns a value of true if the end-of-file marker is found and false otherwise.

Example:

```
$f = fopen($fileName,'r');
while (feof($f)) {
    $line = fgets($f);
    {Statements to process file go here.}
}
fclose($f);
```

(string) fgetc((resource) $fileHandle)

Returns a single character from $fileHandle.

(string) fgets((resource) $fileHandle
                [, (int) $length])

Returns a string of up to $length $-$ 1 bytes from the file pointed to by $fileHandle. If the optional length parameter is not provided, fgets() will read to the end of the line.

Examples:

```
$line = fgets($in, 128);
$theWholeLine = fgets($in);
```

Text files created on one system may cause problems when using fgets() on a different system. UNIX-based files use only a single character, \n, as a line terminator. Windows systems use two characters, \r\n, as a line terminator. As a result, it is possible that fgets() used in a script running on a Windows computer may not properly detect end-of-line marks in a file created on a UNIX system.

$a = file((string) $filename)

Reads an entire file into array $a. When $filename refers to a text file, each line in the file becomes an array element.

Example:

For this data file:

```
Site Lat Lon
brooks 40.01 -75.99
europe 50.5 5.3
south -30 88
farsouth -79 -167
```

this code

```php
<?php
  $a=file("LatLon.dat");
  var_dump($a);
?>
```

produces this output:

array(5) { [0]=> string(14) "Site Lat Lon " [1]=> string(21) "brooks 40.01 –75.99 " [2]=> string(17) "europe 50.5 5.3 " [3]=> string(14) "south –30 88 " [4]=> string(17) "farsouth –79 –167" }

```
(int) fprintf( (resource) $fileHandle,
                (string) {format string}
   [, {one or more arguments to be displayed, comma-separated}])
```

Writes a text string according to the format conversion specifier string, to the file pointed to by $fileHandle. The format type specifiers should match the data type of the arguments.

fprintf() returns an integer value equal to the number of characters written to $fileHandle. Typically, the return value is not needed. The format string is usually specified as a string literal, but it may be assigned to a variable prior to calling fprintf(). This capability allows for script-controlled formatting.

Examples:

```
fprintf($out, "Here is some output.\n");
  // Writes the text into the file.
```

```
$formatString = "%f, %f, %f\n";
fprintf($out,$formatString,$A,$B,$C);
// comma-delimited output
```

The ability to include commas in the output format string means
that it is easy to create a comma-delimited file that can be opened directly
in a spreadsheet. For Excel, these files typically have a .csv extension.
On Windows systems, lines written to a text file, but not necessarily to a
.csv file that will be opened in a spreadsheet, should be terminated with
\r\n rather than just \n. See printf(), below, for more examples of
how to use format specifiers to control the appearance of output.

(string) fread(*(resource) $fileHandle*, *(int) $length*)

Reads up to *$length* bytes from *$fileHandle*, up to 8192 bytes, and
returns the result in a string.

(mixed) fscanf(*(resource) $fileHandle*,
                 $formatString [, *(mixed) $var…*])

Reads a line of text from a file and parses input according to a specified
format string. Without optional *$var* parameters, output is used to create
an array, the elements of which are determined by the format string. If
$var parameters are included, fscanf() returns the number of
parameters parsed. fscanf() will not read past the end-of-line mark if
more format specifiers are provided than there are values in the line. Any
white-space character in the format string matches any whitespace in the
input stream. For example, a tab escape character (\t) in the format string
can match a space character in the input stream.

(int) printf(*(string) $formatString*
              [, *(mixed) $var…*])

Displays a text string according to the format conversion specifier, to the
open window. printf() returns the number of characters written.
Typically, the return value is not needed. *$formatString* is usually
given as a string literal, but it may be assigned to a variable prior to

calling `printf()`, a capability that allows for script-controlled output formatting.

Examples for `fscanf()` and `printf()`:

Document 4.7 uses input file `dateTime.txt`:

```
01/14/2007 17:33:01
02/28/2007 09:15:00
```

Document 4.7 (dateTime.php)

```php
<?php
  $in=fopen("dateTime.txt","r");
  while (!feof($in)) {
    fscanf($in, "%d/%d/%d
%d:%d:%d", $day, $month, $year, $hour, $min, $sec);
      printf("%'02d/%'02d/%4d %'02d:%'02d:%'02d<br
/>", $day, $month, $year, $hour, $min, $sec);
  }
?>
```

```
01/14/2007 17:33:01
02/28/2007 09:15:00
```

Document 4.8 (`formatExample.php`)

```php
<?php
  $a=67;
  $b=.000717;
  $c=-67;
  $d=83.17;
  $e="Display a string.";
  printf("\t%c\n\r%e\n\r%f",$a,$b,$b);
  // no line feeds!
  printf("<br /><br />%s<br />%e<br />%f
          <br />",$a,$b,$b);
  printf("<br />%d %u<br />",$a,$a);
  printf("<br />%d %u<br />",$c,$c);
  // note effect of %u!
  printf("<br />He said, \"Let's go!\"<br />");
  printf("<br />Your discount is \$%'012.2f<br
/>",$d);
```

```
  printf("<br />%'x26s<br />",$e);
?>
```

```
C 7.17000e-4 0.000717

67
7.17000e-4
0.000717

67 67

-67 4294967229

He said, "Let's go!" .

Your discount is
$000000083.17

xxxxxxxxxDisplay a string.
```

Note that printf() ignores the \n, \r, and \t characters when it displays results in your browser window because HTML ignores "white space." This explains the presence of the <br /> tags in the format string. However, fprintf() properly interprets these characters when printing to a file.

(mixed) print_r(*(mixed)* $expression [, *(bool)* $return])

Displays information about $expression, often an array, in a readable format. Setting $return to true copies the output into a variable rather than displaying it.

Example:

```php
<?php
$cars = array("VW", "GM", "BMW", "Saab");
print_r($cars);
$result = print_r($cars,true);
printf("<br />%s",$result);
?>
```

```
array ( [0] => VW [1] => GM [2] => BMW [3] => Saab )
array ( [0] => VW [1] => GM [2] => BMW [3] => Saab )
```

(string) sprintf(*(string)* $formatString
                [, *(mixed)* $var...])

Returns a string built according to the format specifier and optional arguments. $formatString is usually given as a string literal, but may be assigned to a variable prior to calling printf(), a capability that allows for script-controlled formatting.

(mixed) sscanf($line,
                $formatString [, *(mixed)* $var...])

Reads a line of text from a file and parses input according to a specified format string. Without optional $var parameters, output is used to create an array, the elements of which are determined by the format string. If $var parameters are included, sscanf() returns the number of parameters parsed. sscanf() will not read past the end of line if more format specifiers are provided than there are values in the line. Any whitespace character in the format string matches any white space in the input stream. For example, a tab escape character (\t) in the format string can match a space character in the input stream.

(int) vprintf(*(string)* *{format string}*, *(array)* $a)

Displays a string built from the arguments of array $a, formatted according to the specified format string.

Example for sprintf() and vprintf():

Document 4.9 (arrayDisplay.php);

```php
<?php
$a = array("VW", 17.3, "GM", 44, "BMW");
print_r($cars);
$result = print_r($cars,true);
printf("<br />%s",$result);
vprintf("<br />%s, %f, %s, %u, %s",$a);
$result = sprintf("<br />%s, %f, %s, %um %s",
        $a[0],$a[1],$a[2],$a[3],$a[4]);
echo '<br />' . $result;
?>
```

Example:

VW, 17.300000, GM, 44, BMW

VW, 17.300000, GM, 44, BMW

String Handling Functions

(string) chr(*(int) $ascii)*

(int) ord(*(string) $string)*

chr() and ord() are complementary functions. chr() returns the single-character string corresponding to the *$ascii* value. ord() returns the base-10 value of the first character of *$string*. Appendix 2 contains a list of the 256 standard ASCII codes (base 10, 0–255) and their character representations for Windows computers. The lowercase alphabet starts at ASCII (base-10) 97 and the uppercase alphabet starts at ASCII 65. Nearly all ASCII characters can be displayed and printed by using their ASCII codes.

(int) strcasecmp(*(string) $s1, (string) $s2)*

Performs a case-insensitive comparison of *$s1* and *$s2*.
     strcasecmp() returns 0 if *$s1* and *$s2* are identical, an integer less than 0 if *$s1* is less than *$s2* (in the lexical sense), and an integer value greater than 0 if *$s1* is greater than *$s2*.

Examples:
```
strcasecmp(("Dave","David"); // returns -4
strcasecmp("DAVID","david"); // returns 0
```

(int) strcmp(*(string) $s1, (string) $s2)*

Performs a case-sensitive comparison of *$s1* and *$s2*.

strcmp() returns 0 if $s1 and $s2 are identical, an integer value less than 0 if $s1 is less than $s2 (in the lexical sense), and an integer value greater than 0 if $s1 is greater than $s2.

Example:

```
strcmp("david","DAVID"); // returns 1
```

```
(int) strlen($s);
```

Returns the length of string $s.

```
(int) strncasecmp( (string) $s1, (string) $s2, (int)
$n_char)
```

Perform a case-insensitive comparison on the first $n_char characters of $s1 and $s2.

strncasecmp() returns 0 if $s1 and $s2 are identical, an integer value less than 0 if $s1 is less than $s2 (in the lexical sense), and an integer value greater than 0 if $s1 is greater than $s2.

Examples:

```
strncasecmp("Dave","David", 3); // returns 0
strncasecmp(("Dave","David", 4); // returns -4
$len = min(strlen("Dave"),strlen("David"));
strncasecmp("Dave","David",$len);
// compares number of characters contained in shorter
string parameter and returns -4
```

```
(int) strncmp( (string) $s1, (string) $s2, (int)
$n_char)
```

Performs a case-sensitive comparison on the first $n_char characters of $s1 and $s2.

strncmp() returns 0 if $s1 and $s2 are identical, an integer value less than 0 if $s1 is less than $s2 (in the lexical sense), and an integer value greater than 0 if $s1 is greater than $s2.

Examples:

```
strncmp("Dave","David", 3); // returns 0
$len = min(strlen("Dave"),strlen("David"));
strncmp("Dave","David",$len);
// compares number of characters contained in shorter
string parameter and returns -1
```

```
(string) strtolower( (string) $s)
(string) strtoupper( (string) $s)
```

strtolower() converts the alphabetic characters in $s to lowercase.
strtoupper() converts the alphabetic characters in $s to uppercase.

4.4.2 Math Constants and Functions

PHP's math functions return integer or floating-point results, with a system-dependent precision that is often about 14 digits for floating-point numbers. The precision may vary from system to system, but it should be sufficient for all but the most specialized calculations. There are also several pre-defined mathematical constants, all of which are floating-point numbers. Constants and functions are built into PHP, with no need for external libraries.[1] Trigonometric functions always assume input parameters in radians or produce angle outputs in radians. Data types are shown in parentheses, for example, *(float)*. Optional arguments are enclosed in square brackets.

Constants and built-in math functions are listed in Table 4.5. In this table, "x" (and other arguments, in some cases) always represents a variable of the appropriate type, even though they are shown without the $ symbol.

Table 4.5 Math constants and functions

Named constants	Description
M_1_PI	$1/\pi$
M_2_PI	$2/\pi$
M_2_SQRTPI	$2/(\pi^{1/2})$

[1] C, for example, requires that a math library be linked to a source code file before it is compiled.

`M_E`	Base of the natural logarithm, $e = 2.71828\ldots$
`M_EULER`	Euler's constant [1] $= 0.577215665\ldots$
`M_LN2`	Natural logarithm of $2 = 0.693147\ldots$
`M_LN10`	Natural logarithm of $10 = 2.302585\ldots$
`M_LNPI`	Natural logarithm of $\pi = 1.1447299\ldots$
`M_LOG2E`	Log to the base 2 of $e = 1.442695\ldots$
`M_LOG10E`	Log to the base 10 of $e = 0.434294\ldots$
`M_PI`	$\pi = 3.1415927\ldots$
`M_PI_2`	$\pi/2 = 1.5707963\ldots$
`M_PI_4`	$\pi/4 = 0.7853981\ldots$
`M_SQRT1_2`	$1/(2^{1/2}) = 0.7071067\ldots$
`M_SQRT2`	$2^{1/2} = 1.4142136\ldots$
`M_SQRT3`	$3^{1/2} = 1.7320508\ldots$
`M_SQRTPI`	$\pi^{1/2} = 1.7724539\ldots$
Functions	**Returns**
`(number)abs((number)x)`	Absolute value of x, a floating-point or integer number depending on x
`(float)acos((float)x)`	Inverse cosine of x, $\pm\pi$, for $-1 \leq x \leq 1$
`(float)acosh((float)x)`	Inverse hyperbolic cosine of x [2]
`(float)asin((float)x)`	Inverse sine of x, $\pm\pi/2$, for $-1 \leq x \leq 1$
`(float)asinh((float)x)`	Inverse hyperbolic cosine of x [2]
`(float)atan((float)x)`	Arc tangent of x, $\pm\pi/2$, for $-\infty < x < \infty$ (compare with `Math.atan2(y,x)`)
`(float)atan2((float)y, (float)x)`	Inverse tangent of angle between x-axis and the point (x, y), 0–2π, measured counterclockwise

`(float)atanh((float)x)`	Inverse hyperbolic tangent of x [2]
`(float)ceil((number)x)`	Smallest whole number (still type `float`) greater than or equal to x
`(float)cos((float)x)`	Cosine of x, ±1
`(float)cos((float)x)`	Hyperbolic cosine of x
`(float)deg2rad((float)x)`	Convert x in degrees to radians
`(float)exp((float)x)`	e to the x power (e^x)
`(float)floor((float)x)`	Greatest whole number (still type `float`) less than or equal to x
`(float)fmod((float)x, (float)y)`	Floating-point remainder of x/y
`(float)log((float)x[, (float)b])`	Logarithm of x, to base e unless optional base argument b is included, x>0
`(float)getrandmax((void))`	Max value returned by call to `rand()`
`(float)log10((float)x)`	Base-10 logarithm of x
`(mixed)max((mixed)x, (mixed)y)` `(mixed)max((array)x)`	Greater of x or y, or maximum value in an array
`(mixed)min((mixed)x, (mixed)y)` `(mixed)min((array)x)`	Lesser of x or y, or minimum value in an array
`(float)pi()`	Returns value of π, identical to `M_PI`
`(number)pow((number)x, (number)y)`	x to the y power (x^y). Returns an integer whenever possible
`(float)rad2deg((float)x)`	Convert radian value x to degrees
`(int)rand()` `(int)rand([(int)min, (int)max])`	Random real number in the range 0–`RAND_MAX`, optionally between `min` and `max`

(float) round (*(float)* x [, *(float)* p])	x rounded to specified precision (p digits after decimal point), or to whole number without argument p
(float) sin (*(float)* x)	Sine of x
(float) sinh (*(float)* x)	Hyperbolic sine of x
(float) sqrt (*(float)* x)	Square root of x
(float) srand ([*(int)* seed])	Seeds random number generator, optionally with specified integer seed.
(float) tan (*(float)* x)	Tangent of x, ±∞
(float) tanh (*(float)* x)	Hyperbolic tangent of x

[1]Euler's constant is the limit as n→∞ of (1 + 1/2 + 1/3 + ... + 1/n) − ln(n).
[2]Not implemented in all versions of PHP.

4.4.3 Array Functions and Constructs

```
(array) array([(mixed) {comma-separated list of arguments}])
// This is a construct, not a function
```

Creates an array from specified arguments. It is not necessary to provide arguments when the array is created.

```
(array) array_keys((array) $a)
```

Returns an array containing the keys of the $a array.

```
(mixed) array_pop ((array) $a )
```

Treats $a as a stack and removes and returns the last (newest) element of $a, automatically shortening $a by one element. A value of NULL will be returned if the array is already empty. This functions resets the array pointer to the beginning of the array after the element is removed.

Example (for array() and array_pop()):

Document 4.10 (arrayPop.php)

```php
<?php
$stack = array("orange", "banana", "apple", "lemon");
$fruit=array_pop($stack);
print_r($stack);
?>
```

```
Array ( [0] => orange [1] => banana [2] => apple)
```

The variable $fruit will be assigned a value of *lemon*.

(int) array_push(*(array) $a, (mixed) $var* [, *(mixed)*...])

Treats $a as a stack, and pushes the passed variable(s) onto the end of $a. The length of $a increases by the number of variables pushed. Returns the number of elements in the array after the "push."

Example:

Document 4.11 (arrayPush.php)

```php
<?php
$stack = array("red", "grn");
$n = array_push($stack, "blu", "wh");
print_r($stack);
$stack[] = "blk";
printf("<br />%u<br />",$n);
print_r($stack);
printf("<br />%u<br />",sizeof($stack));
?>
```

```
array ( [0] => red [1] => grn [2] => blu [3] => wh )
4
array ( [0] => red [1] => grn [2] => blu
        [3] => wh [4] => blk )
5
```

The shaded line in Document 4.11 shows that a new variable can be "pushed" onto the end of an array simply by assigning a new element to the array. Because this avoids whatever overhead might be associated with a function call, and it is shorter to write, it might make sense to use array_push() only when you wish to add multiple new values at the same time.

(mixed) `array_shift((array) $a)`

Removvs the first element of *$a* (the "oldest" element) and returns it, then shortens *$a* by one element and moves everything down one position. Numerical keys will be reset to start at 0. Literal keys are unchanged. `array_shift()` is used to remove the oldest element from an array treated as a queue. It resets the array pointer to element 0 after it is used.

Example:

Document 4.12 (`arrayShift.php`)

```php
<?php
$queue = array("orange", "banana", "raspberry", "mango");
print_r($queue);
$rottenFruit = array_shift($queue);
echo '<br />' . $rottenFruit;
echo '<br />' . count($queue);
?>
```

```
Array ( [0] => orange [1] => banana [2] => raspberry
[3] => mango )
orange
3
```

(int) `array_unshift((array) $a),`
                    (mixed) $var [, (mixed)...])

Adds one or more elements to the "front" of the array (the "old" end). The entire list is inserted in order, so the first item in the list to be added is the first element in the modified array. Numerical keys are reset to start at 0. Literal keys are unchanged.

Example:

Document 4.13 (`arrayUnshift.php`)

```php
<?php
$a = array("orange", "banana", "raspberry", "mango");
print_r($a);
array_unshift($a,"papaya","mangosteen");
echo '<br />' . count($a) . '<br />';
print_r($a);
?>
```

```
Array ( [0] => orange [1] => banana [2] => raspberry
      [3] => mango )
6
Array ([0] => papaya [1] => mangosteen [2] => orange
      [3] => banana [4] => raspberry [5] => mango )
```

(int) count(*(mixed)* $a [, $mode])

(int) sizeof(*(mixed)* $a [, $mode])

count() and sizeof() are equivalent. They return the number of elements in the array $a. If the value of $mode if it is not specified, its default value is 0. Setting $mode to 1 or to COUNT_RECURSIVE will count elements recursively in a multidimensional array.

The "recursive count" might not do what you expect. In a two-dimensional array with 5 "rows" and 4 "columns" (refer to Document 3.4, two-D.php), the recursive count option counts 5×4 rows, and then 5 rows again, and returns a value of 25. The number of elements in this two-dimensional array is not 25, but 25 − 5 = 20.

4.4.4 Miscellaneous Functions and Constructs

break [*(int)* $n]

Exits the current conditional or count-controlled loop structure. An optional argument following break (not in parentheses) specifies the number of nested structures to be exited.

(bool) copy(*(string)* $source,*(string)* $destination)

(bool) rename(*(string)* $source,*(string)* $destination)

copy() copies the $source file to the $destination file. rename() renames the $source file to $destination. They both return true or false depending on whether they were successful.

```
die ( [ (string) $status])

die ( [ (int) $status])

exit ([ (string) $status])

exit ([ (int) $status])
```

Equivalent functions to exit a script. If the argument is a string, it will be
printed on exit. An integer argument, in the range 0–254, is available for
use as an exit error code in other applications, but it is not printed.

```
(array) explode ( (string) $delimiter, (string) $s,
                        [ (int) $n])
(string) implode ( (string) $delimiter, (array) $a )
```

explode () returns an array of strings consisting of substrings of the
string $s, in which the substrings are separated by the $delimiter.
When $n is present, explode () will build array elements from the first
$n values, with the last element containing the remainder of the string.
The delimiter must match the file contents exactly. For example, a " "
(single space) delimiter implies that the values are separated by one and
only one space. In a file with numerical values, the elements of the
returned array can be treated as numbers in subsequent code.

  implode () returns all elements of $a as a concatenated string,
with the elements separated by $delimiter.

Example:

Using this data file, LatLon.dat

```
Site Lat Lon
brooks 40.01 -75.99
europe 50.5 5.3
south -30 88
farsouth -79 -167
```

Document 4.14 (ExplodeArray.php)

```php
<?php
  $a=file("LatLon.dat");
  var_dump($a);
  echo "<br />";
  for ($i=1; $i<sizeof($a); $i++) {
    list($s,$la,$lo)=explode(" ",$a[$i]);
      echo $s.", ".$la.", ".$lo."<br />";
  }


  foreach ($a as $s) {
    list($site,$Lat,$Lon)=explode(" ",$s);
      echo $site.", ".$Lat.", ".$Lon."<br />";
  }
?>
```

array(5) { [0]=> string(14) "Site Lat Lon " [1]=> string(21) "brooks 40.01 –75.99 " [2]=> string(17) "europe 50.5 5.3 " [3]=> string(14) "south –30 88 " [4]=> string(17) "farsouth –79 –167" }
brooks, 40.01, –75.99
europe, 50.5, 5.3
south, –30, 88
farsouth, –79, –167
Site, Lat, Lon
brooks, 40.01, –75.99
europe, 50.5, 5.3
south, –30, 88
farsouth, –79, –167

(int) strtotime(*(string) $time*)

Converts a date and time description, in any common format, into the number of seconds from January 1, 1970, 00:00:00 GMT. For dates specified in xx/xx/xx or xx/xx/xxxx format, strtotime() assumes the U.S. custom of supplying dates as mm/dd/yy or mm/dd/yyyy. (The European custom is to specify dates as dd/mm/yy or dd/mm/yyyy.) strtotime() can be used to determine whether a date comes before or after another date.

Example:

echo strtotime("12/04/2007"); yields the result 1196744400

```
(void) var_dump ( (mixed) $var1 [, (mixed) $var2])
```

Displays structured information about one or more variables.

Example:

Document 4.15 (varDump.php)

```php
<?php
$a = array('david','apple','Xena','Sue');
$b = array();
list($b[0],$b[1],$b[2],$b[3]) = $a;
var_dump($b);
?>
```

```
array(4) {
[3]=> string(3) "Sue" [2]=> string(4) "Xena" [1]=>
string(5) "apple"
[0]=> string(5) "david" }
```

```
(void) list( (mixed) {arguments} ) = $array
//a construct, not a function
```

Assigns contents of an array to several variables.

Examples:

Document 4.16 (arrayList.php)

```php
<?php
$stuff = array('I','love','PHP.');
list($who,$do_what,$to_what) = $stuff;
echo "$who $do_what $to_what" . "<br />";
list($who, , $to_what) = $stuff;
echo "$who $to_what<br />";
$a = array('david','apple','Xena','Sue');
$b = array();
list($b[0],$b[1],$b[2],$b[3]) = $a;
var_dump($b);
echo "<br />Access with for... loop.<br />";
for ($i=0; $i<count($b); $i++) echo $b[$i] . "<br />";
echo "Access with foreach... loop.<br />";
foreach ($b as $key => $x) echo "a[" . $key . "] = " . $x .
"<br />";
?>
```

```
I love PHP.
I PHP.
array(4) { [3]=> string(3) "Sue" [2]=> string(4) "Xena"
[1]=> string(5) "apple" [0]=> string(5) "david" }
Access with for... loop.
david
apple
Xena
Sue
Access with foreach... loop.
a[3] = Sue
a[2] = Xena
a[1] = apple
a[0] = david
```

Note that with scalar, named variables, as in

```php
$stuff = array('I','love','PHP.');
list($who,$do_what,$to_what) = $stuff;
```

the result is what you expect. However, if the target of the list operation is an array, as in

```php
$a = array('david','apple','Xena','Sue');
$b = array();
list($b[0],$b[1],$b[2],$b[3]) = $a;
```

then the output shows that the order of the keys is reversed. That is, the first key for the $b array is 3 and not 0. If you use a for... loop with the numerical indices, you can still get elements printed in the same left-to-right order in which they are defined in $a, but if you use a foreach... loop to display the contents of $b, the order will be reversed.

5 Using PHP from a Command Line

Chapter 5 gives a brief introduction to using PHP from a command line. This capability does not require that PHP run on a server and it allows user input from the keyboard while a script is executing.

5.1 The Command Line Environment

Throughout this book, the typical model for using PHP has been to create an HTML document that serves as an interface to pass form field values as input to a PHP application running on a local or remote server. Those values are automatically sent to the $_POST[] array. Some of the shorter PHP code examples—those that do not require user input—run as stand-alone applications on a server. For example, some of the examples in Chapter 3 use "hard-coded" array elements just to illustrate some syntax for processing arrays.

In that HTML/server PHP implementation, there was no provision for entering input from the keyboard. In some cases, it might be convenient to be able to run stand-alone PHP applications with keyboard input. It is possible to do this from a **command line interface (CLI)**. Doing so removes the possibilities for HTML formatting of PHP output in a browser window, so you may find this to be a practical solution only for calculations with simple output requirements.

The first step toward learning how command line PHP works on a local computer is to find where the php.exe program resides. On a local computer, this is probably not the same folder from which you have previously executed PHP applications on your local server. Assume that this file is located in C:\PHP.

Next, create this simple PHP file with a text editor and store it as hello.php in C:\PHP:

```
<?php
  echo "Hello, world!";
?>
```

CLI 5.1 shows a record of a Windows command line session that executes this file. You can type the line as shown or you can type

D.R. Brooks, *Introduction to PHP for Scientists and Engineers*,
doi: 10.1007/978-1-84800-237-1_5, © Springer-Verlag London Limited 2008

`php.exe hello.php`—the `.exe` extension is assumed on Windows computers.

CLI 5.1

```
C:\PHP>php hello.php
Hello, world!
C:\PHP>
```

This is a trivial PHP "application," but it is important because it differs fundamentally from what has been presented in all the previous chapters of this book. This PHP application runs directly from the directory in which the `php.exe` application resides—`C:\PHP` on this computer. In fact, `hello.php` can be executed from any directory that contains a copy of the `php.exe` file. This application did *not* run on a server!

There are several command line options that can be used when a PHP file is executed, but they are not needed for the simple examples shown in this chapter. As always, there are many online sources of more information about using a CLI with PHP.

PHP's command line capabilities make much more sense if you can provide input to a PHP application that actually does something useful. Consider this problem:

> Write a stand-alone application that allows a user to enter an upper and lower limit and then calculates the integral of the normal probability density function,
>
> $$\text{pdf}(x) = \frac{\exp(-x^2/2)}{\sqrt{2\pi}}$$
>
> using those two limits. This function cannot be integrated analytically, so numerical integration is required. There are several ways to integrate functions numerically, but so-called Trapezoidal Rule integration will work well for this problem:
>
> $$\int_{x_a}^{x_b} \text{pdf}(x) \approx \left(\sum_{i=1}^{i=n-1} [f(x_i) + f(x_i + \Delta x)] \right) \frac{\Delta x}{2}$$

Start the code for a CLI application with this short script:

Document 5.1a (pdf_1.php, partial)

```php
<?php
  $a = $_SERVER['argv'];
  print_r($a);
?>
```

In the same way that $_POST[] contains values passed from an HTML document, the 'argv' element of the $_SERVER[] array contains the values passed from a command line. CLI 5.2 shows the execution of this script:

CLI 5.2

```
C:\PHP>php pdf_1.php -.5 .5
Array
(
      [0] => pdf_1.php
      [1] => -.5
      [2] => .5
)
```

Note that the arguments passed to the PHP application through the 'argv' array include the file name of the application itself as the first element. Therefore, the lower and upper limits for the numerical integration are the second and third elements of $a, $a[1] and $a[2]. Document 5.1b shows the complete code for this problem.

Document 5.1b (pdf_1.php)

```php
<?php
  $a = $_SERVER['argv'];
  print_r($a);
  $x1=$a[1]; $x2=$a[2];
  $n=200;
  $sum=0; $dx=($x2-$x1)/$n;
  for ($i=1; $i<=$n; $i++) {
    $x=$x1+($i-1)*$dx;
    $y1=exp(-$x*$x/2)/sqrt(2.*M_PI);
    $x=$x1+$i*$dx;
    $y2=exp(-$x*$x/2)/sqrt(2.*M_PI);
    $sum+=$y1+$y2;
  }
  echo "\n" . $sum*$dx/2.;
?>
```

CLI 5.3

```
C:\PHP>php pdf_1.php -.5 .5
Array
(
    [0] => pdf_1.php
    [1] => -.5
    [2] => .5
)

0.38292418907776
C:\PHP>
```

CLI 5.3 shows a command line session that executes this code. The application expects you to provide the upper and lower integration limits after the PHP file name. No prompts are provided for this information, and it is the user's responsibility to know what needs to be entered. Note that the HTML formatting tags that have been used in previous chapters—
 to produce a line break, for example—will not work in this environment. Instead, the final echo statement in Document 5.1b contains a line feed escape character, \n.

In general, it would be more helpful to be able to provide prompts to the user about required input from within a PHP application being executed from the CLI. Document 5.2 shows another approach to evaluating the normal probability distribution function which prompts user input from the keyboard, to be entered while the script is executing.

Document 5.2 (pdf_2.php)

```php
<?php
  echo "\nGive lower and upper limits for evaluating
pdf,\nseparated by a space: ";
  fscanf(STDIN,"%f %f",$x1,$x2);
  echo $x1 . ", " . $x2;
  $n=200;
  $sum=0; $dx=($x2-$x1)/$n;
  for ($i=1; $i<=$n; $i++) {
    $x=$x1+($i-1)*$dx;
    $y1=exp(-$x*$x/2)/sqrt(2.*M_PI);
    $x=$x1+$i*$dx;
    $y2=exp(-$x*$x/2)/sqrt(2.*M_PI);
    $sum+=$y1+$y2;
  }
  echo "\n" . $sum*$dx/2.;
?>
```

CLI 5.4

```
C:\PHP>php pdf_2.php

Give lower and upper limits for evaluating pdf,
separated by a space: -3 3
-3, 3
0.99729820978444
```

Document 5.2 uses the `fscanf()` function. But, instead of using a file handle as the input resource, `fscanf()` uses the reserved name `STDIN` (which must be written in uppercase letters), which identifies the keyboard as the input resource. The keyboard can be designated as the input resource for any of the other input functions that require a resource identifier, such as `fgets()` and `fread()`.

It is possible to write PHP applications that will execute either from a CLI or on a server through an HTML document. Document 5.3a provides an HTML interface and 5.3b is a PHP application that will work either on a server or as a stand-alone CLI application.

Document 5.3a (`pdf_3.htm`)

```
<html>
<head>
<title>Integrate the normal probability density
function</title>
</head>
<body>
<h3>Evaluate the normal probability density function</h3>
<form method="post" action="pdf_3.php">
x1: <input type="text" name="x1" value="-0.5" /><br />
x2: <input type="text" name="x2" value=".5"   /><br />
<input type="submit" value="Click to evaluate." />
</form>
</body>
</html>
```

Document 5.3b (`pdf_3.php`)

```
<?php
  if ($_SERVER['argc'] > 0) {
    $a = $_SERVER['argv'];
    print_r($a);
    $x1=$a[1]; $x2=$a[2];
  }
  else {
    $x1=$_POST['x1'];
```

```
    $x2=$_POST['x2'];
    echo $x1 . ", " . $x2 . "<br />";
 }
 $n=200;
 $sum=0; $dx=($x2-$x1)/$n;
 for ($i=1; $i<=$n; $i++) {
   $x=$x1+($i-1)*$dx;
     $y1=exp(-$x*$x/2)/sqrt(2.*M_PI);
     $x=$x1+$i*$dx;
     $y2=exp(-$x*$x/2)/sqrt(2.*M_PI);
     $sum+=$y1+$y2;
 }
 echo $sum*$dx/2.;
?>
```

When Document 5.3b is run from a server, the output looks like this:

```
-0.5, .5
0.38292418907776
```

When Document 5.3b is run from a CLI, the output looks like it did for CLI 5.3.

In Document 5.3b, the `'argc'` element of `$_SERVER[]` contains the number of command line parameters passed to the script when it is executed in a CLI. If this value is 0, then the alternate path is executed to retrieve the values passed from Document 5.3a.

5.2 Is a Command Line Interface Useful?

The capabilities introduced in this chapter for passing arguments from a command line and accepting user input typed at a keyboard should be very familiar to C programmers, an observation that most readers of this book may find totally irrelevant. Whether you find using a CLI for some PHP applications useful or a giant leap backwards into the long-gone and best forgotten days of text-based computing may depend on your previous programming experience and quite possibly your age!

There is no doubt that a text-based CLI is primitive by the standards of today's graphical user interfaces (GUIs), but it still has its place for some kinds of applications. Once programmers started using PHP for web applications, they realized that if scripts could be executed from a CLI, it would be useful for many of the offline system-related tasks that are required to maintain a large web site. Unlike server-based PHP, CLI-based PHP scripts do not require close attention to file access privileges, which can be a major time saver for a web site manager. Also, the programming overhead for these kinds of tasks can be much lower with a simple text-based interface than it would be for more modern GUIs. Finally, CLI scripts run very quickly in this text-based environment because they do not depend on much larger and more complex GUI applications. So, CLI-based PHP quickly became very popular with professional programmers.

For the casual programmer, the arguments favoring the use of a CLI are less compelling. However, it is worth remembering that when PHP scripts run from a CLI, they are completely portable because they do not require a server. You can, for example, store such applications on a USB pen drive along with the `php.exe` file and run them anywhere. When you develop your own PHP applications, it may be worth considering whether they can or should be made CLI-compatible, as was done in Document 5.3b.

Appendices

A.1 List of HTML and PHP Document Examples

Document and Name		Page
1.1	getCalib.htm	3
1.1	getCalib.php	5
1.3	helloWorld.php	8
1.4	PHPInfo.php	9
1.5	writeCalib.php	15
1.6a	quadrat.htm	18
1.6b	quadrat.php	19
1.7a	WeatherReport.htm	20
1.7b	WeatherReport.php	21
2.1	PWcalc2.htm	25
2.2	PWcalc3.htm	30
2.3	PWcalc3.php	33
2.4	circleStuff.php	39
2.5	windspd.php	41
2.6a	getMass.htm	44
2.6b	getMass.php	47
3.1	keyedArray.php	50
3.2	ConsecutiveKeyArray.php	51
3.3	base_1Array.php	52
3.4	two-D.php	53
3.5	sort1.php	54
3.6	sort2.php	55
3.7	sort3.php	56
3.8	StacksAndQueues.php	57
3.9a	quadrat2.htm	59
3.9b	quadrat2.php	60

3.10a   quadrat3.htm   61
3.11b   quadrat3.php   61
3.11a   CloudObs.htm   63
3.11b   CloudObs.php   65
3.12   histo.php   66
3.13   cardShuffle.php   68
3.14a   DataReport.htm   72
3.14b   DataReport.php   72

4.1   conditionalExecution.php   77
4.2   daysInMonth.php   78
4.3   countdown.php   80
4.4   loopExamples.php   80
4.5   foreach.php   82
4.6   squares.php   83
4.7   dateTime.php   91
4.8   formatExample.php   91
4.9   arrayDisplay.php   93
4.10   arrayPop.php   99
4.11   arrayPush.php   100
4.12   arrayShift.php   101
4.13   arrayUnshift.php   101
4.14   explodeArray.php   103
4.15   varDump.php   105
4.15   arrayList.php   105

CLI5.1 (Running hello.php from a command line.)   108
5.1a   pdf_1.php (partial)   109
5.1b   pdf_1.php (complete)   109
CLI5.3 (Running pdf_1.php from a command line.) 110
5.2   pdf_2.php   110
CLI5.4 (Running pdf_2.php from a command line.) 111
5.3a   pdf_3.htm   111
5.3b   pdf_3.php   111

A.2 ASCII Characters for Windows PCs

The first 127 ASCII character codes are standardized and the remaining characters are system-dependent. The values shown are for Windows-based PCs. These characters can be displayed from a Windows computer keyboard by pressing and holding the Alt key and pressing the corresponding base-10 (Dec) code on the numerical keypad ("locked" with the NumLock key).

Dec	Hex		Dec	Hex		Dec	Hex	
0	0	(1)	30	1E	▲	61	3D	=
1	1	☺	31	1F	▼	62	3E	>
2	2	☻	32	20	(2)	63	3F	?
3	3	♥	33	21	!	64	40	@
4	4	♦	34	22	"	65	41	A
5	5	♣	35	23	#	66	42	B
6	6	♠	36	24	$	67	43	C
7	7	•	37	25	%	68	44	D
8	8	◘	38	26	&	69	45	E
9	9	○	39	27	'	70	46	F
10	A	◙	40	28	(	71	47	G
11	B	♂	41	29	)	72	48	H
12	C	♀	42	2A	*	73	49	I
13	D	♪	43	2B	+	74	4A	J
14	E	♫	44	2C	,	75	4B	K
15	F	☼	45	2D	-	76	4C	L
16	10	►	46	2E	.	77	4D	M
17	11	◄	47	2F	/	78	4E	N
18	12	↕	48	30	0	79	4F	O
19	13	‼	49	31	1	80	50	P
20	14	¶	50	32	2	81	51	Q
21	15	§	51	33	3	82	52	P
22	16	▬	52	34	4	83	53	S
23	17	↨	53	35	5	84	54	T
24	18	↑	54	36	6	85	55	U
25	19	↓	55	37	7	86	56	V
26	1A	→	56	38	8	87	57	W
27	1B	←	57	39	9	88	58	X
28	1C	∟	58	3A	:	89	59	Y
29	1D	↔	59	3B	;	90	5A	Z
			60	3C	<	91	5B	[

92	5C	\	134	86	å	176	B0	
93	5D	]	135	87	ç	177	B1	
94	5E	^	136	88	ê	178	B2	
95	5F	_	137	89	ë	179	B3	
96	60	`	138	8A	è	180	B4	
97	61	a	139	8B	ï	181	B5	
98	62	b	140	8C	î	182	B6	
99	63	c	141	8D	ì	183	B7	
100	64	d	142	8E	Ä	184	B8	
101	65	e	143	8F	Å	185	B9	
102	66	f	144	90	É	186	BA	
103	67	g	145	91	æ	187	BB	
104	68	h	146	92	Æ	188	BC	
105	69	i	147	93	ô	189	BD	
106	6A	j	148	94	ö	190	BE	
107	6B	k	149	95	ò	191	BF	
108	6C	l	150	96	û	192	C0	
109	6D	m	151	97	ù	193	C1	
110	6E	n	152	98	ÿ	194	C2	
111	6F	o	153	99	Ö	195	C3	
112	70	p	154	9A	Ü	196	C4	
113	71	q	155	9B	¢	197	C5	
114	72	r	156	9C	£	198	C6	
115	73	s	157	9D	¥	199	C7	
116	74	t	158	9E	Pts	200	C8	
117	75	u	159	9F	f	201	C9	
118	76	v	160	A0	á	202	CA	
119	77	w	161	A1	í	203	CB	
120	78	x	162	A2	ó	204	CC	
121	79	y	163	A3	ú	205	CD	
122	7A	z	164	A4	ñ	206	CE	
123	7B	{	165	A5	Ñ	207	CF	
124	7C	|	166	A6	a	208	D0	
125	7D	|	167	A7	o	209	D1	
126	7E	}	168	A8	¿	210	D2	
127	7F	△	169	A9	⌐	211	D3	
128	80[3]	Ç	170	AA	¬	212	D4	
129	81	ü	171	AB	½	213	D5	
130	82	é	172	AC	¼	214	D6	
131	83	â	173	AD	¡	215	D7	
132	84	ä	174	AE	«	216	D8	
133	85	à	175	AF	»	217	D9	

218	DA	⌐	231	E7	τ	244	F4	⌠
219	DB	■	232	E8	Φ	245	F5	⌡
220	DC	▄	233	E9	Θ	246	F6	÷
221	DD	▌	234	EA	Ω	247	F7	≈
222	DE	▐	235	EB	δ	248	F8	°
223	DF	▀	236	EC	∞	249	F9	·
224	E0	α	237	ED	φ	250	FA	·
225	E1	ß	238	EE	ε	251	FB	√
226	E2	Γ	239	EF	∩	252	FC	ⁿ
227	E3	π	240	F0	≡	253	FD	²
228	E4	Σ	241	F1	±	254	FE	■
229	E5	σ	242	F2	≥	255	FF	(4)
230	E6	μ	243	F3	≤			

[1] ASCII 0 is a null character.

[2] ASCII 32 is a space (as produced by pressing the space bar on your keyboard).

[3] Because the Euro did not exist when the ASCII character sequence was standardized, its symbol, €, does not have a representation in the standard sequence (although it is available as a special character for many fonts in Microsoft Word, for example). On some European computer systems, it may take the place of Ç, the character for ASCII code 128.

[4] ASCII 255 is a blank character.

Exercises

These exercises, while not keyed to specific chapters, are nonetheless presented roughly in order relative to the material presented in Chapters 1 through 3, and 5. (Chapter 4 is a summary of PHP elements and contains just syntax examples rather than new applications.)

1. Rewrite Document 2.5 (`windspd.php`) so that it uses the `$array = file($filename)` function to copy all the wind speed data into an array. Each line in the file will become an array element, and the `explode()` function can then be used to access the data. This approach can be used to eliminate the long format specifier string required in Document 2.5. To use the `explode()` function, the only requirement is that you know exactly how the values in the file are separated. In the sample file shown in the problem statement for Document 2.5, there is a header line with values separated by a space. Each wind speed value is followed by a comma and a space. Does it matter if the delimiter given in the `explode()` function is `", "` or `","`? The documentation for the `explode()` function says that it "returns an array of strings consisting of substrings of the string" specified as a parameter. Does it matter that the contents of the arrays returned by `explode()` should be treated as numbers and not strings?

2. Create a file of names and densities of various materials. Write an HTML/PHP application that will display all materials and densities for which the density is greater than or less than some value specified in the HTML document. (Use a radio button to select.)

3. Create a file of unit conversions that lists a "from" unit, a "to" unit, and the number by which the "from" unit must be multiplied to get the value of the "to" unit. For example, to convert from feet to yards, multiply by 0.33333. Write an HTML/PHP application that will allow the user to specify a name and value for a "from" unit and the name of a "to" unit, and will display the equivalent value for the "to" unit.

One problem with such an application is that your HTML document will not "know" which units are included in the data file, and that file can be accessed only through the PHP application. You do not want to "hard code" all possible unit names into your HTML document. There is no simple solution to this problem. The PHP application could search the file entries based on only the first few letters of the unit names passed from the HTML document, which would prevent problems with a user specifying "meters" in the HTML document when the data file contained only "meter."

4. Write a PHP application that uses one or more functions to find the minimum, maximum, mean, median, and standard deviation of numerical values stored in a file. Note that to find the median, the values need to be sorted. For an odd number of sorted values, the median is the middle value. For an even number of values, the median is the average of the two middle numbers.

The mean of a list of n numbers is:

$$m = \sum_{x=1}^{n} x_i$$

and the standard deviation is:

$$s = \sqrt{\frac{\sum_{i=1}^{n} x_i^2 - \left(\sum_{i=1}^{n} x_i\right)^2 / n}{n-1}}$$

5. Simulation studies in science and engineering often require random numbers drawn from a normal ("bell-shaped") distribution rather than from a uniform distribution. By definition, a set of normally distributed numbers should have a mean of 0 and a standard deviation of 1. PHP has a random number function that generates uniformly distributed values in the range [0,1). That is, the generator could produce a value of 0, but it should never produce a value of 1.

There is a simple way to generate a pair of normally distributed numbers x_1 and x_2 (or at least numbers that *look* like they are normally distributed in some statistical sense) from a pair of uniformly distributed numbers u_1 and u_2 in the range (0,1]:

$x_1 = [-2 \, ln(u_1)]^{1/2} \cdot \cos(2\pi u_2)$
$x_2 = [-2 \, ln(u_1)]^{1/2} \cdot \sin(2\pi u_2)$

where $\ln()$ is the natural (base e) logarithm.

Write a PHP application to create a file containing 100 normally distributed values. When you do this in a for… loop, remember that each "trip" through the loop calculates two values, x_1 and x_2, not just one. If you like, you can write a simple HTML interface to specify the number of normally distributed values to be generated.

Because $\ln(0)$ is undefined, your code will have to check every value of u_1 to make sure it is not 0. If it is, replace u_1 with some arbitrary small value. This should happen only rarely, if ever, so it will not bias the statistics of even a fairly small sample. Note that PHP's random number generator is not supposed to produce a value of exactly 1, for which $\ln(1) = 0$, but if it does it will cause no problems with these calculations.

Calculate the mean and standard deviation of the numbers you have generated. You can do this by summing the values of x_i and x_i^2 as you generate them. Then use the formulas given in the previous exercise. The mean and standard deviation of these 100 normally distributed values should be close to 0 and 1, but not exactly equal to these values for this finite set. There are other ways to check whether a set of numbers is really normally distributed, but that is beyond the scope of this problem. It is even possible that the numbers generated with this algorithm would pass such tests even though they are not really randomly normal.

6. Snell's law of refraction relates the angle of incidence θ_i of a beam of light to the angle of refraction θ_r of the beam as it enters a different medium:

$n_i\sin(\theta_i) = n_r\sin(\theta_r)$

The table gives the refractive index for four materials. Assuming that the incident material is always air, create a table that shows incident angles from 10° to 90° in steps of 10°. The angle of refraction corresponding to an incident angle of 90° is the angle beyond which light incident from within the refracting material is reflected back into that medium, rather than exiting into air.

Material	Index of refraction
Air	1.00
Water	1.33
Glass	1.50
Diamond	2.42

7. A circuit containing an inductance of L henrys and a capacitance of C farads has a resonant frequency f given by:

$$f = \frac{1}{2\pi\sqrt{LC}}\,\text{Hz}$$

Write an HTML/PHP application that allows the user to input a range of inductances and capacitances along with a "step size" for each component, and generates a table containing the resonant frequency for each LC pair of values.

For example, the output could generate a table for inductances in the range from 20-100 μh in steps of 20 μh and capacitances from 100-1000 μf in steps of 100 μf. It does not make any difference which of these components are the rows in the table and which are the columns.

8. In a materials testing experiment, samples are given random doses of radiation R every hour. The maximum total radiation exposure R_{max} is specified and the experiment is stopped if the next radiation dose will cause R_{max} to be exceeded. The units for the radiation do not matter for this problem.

Write an HTML document that will provide, as input to a PHP application, the maximum total dose and the maximum individual dose. The PHP application should then generate a table summarizing the random doses delivered to the sample. It could look something like this:

Maximum cumulative radiation = 1000		
Maximum individual dose = 200		
Dose	Amount	Cumulative
1	144	144
2	200	344
3	73	417
4	59	476
5	168	644
6	119	763
7	99	862
8	177	not delivered

9. Modify Document 3.12 (histo.php) so that the code will accommodate any arbitrary range of input values, with those values distributed in the specified number of equal-size bins. For example, the occurrences of numbers from −30 to +30 could be counted in 12 bins of size 5. It should be up to the user to define the number of bins in a way

that makes sense relative to the range of values to be represented by the histogram. The code should save the histogram data in a file.

10. Create a data file containing an unspecified number of values between 0 and 100. Define an array with letter grades as keys:

```
$a = array("A" => 90, "B" => 80, ...);
```

This array defines cutoff points for each letter grade.

      Define another array with the same character keys. The elements of this array should be initialized to 0. Then, when you read through your data file, increment the appropriate grade "box" by 1. (This could be done with multiple if... statements, for example.) When you are finished, display the keys and contents of the second array in a table that shows the number of A's, B's, etc.; for example:

```
A 3
B 7
C 5
D 2
F 1
```

      This is just another version of the histogram problem, but the box limits are defined by the elements of $a, rather than by allocating the values into a specified number of identically sized boxes.

11. Write a PHP application that will read a text file and count the number of occurrences of each letter in the file. (Use the `fgetc()` function to read one character at a time from the file.) Upper- and lowercase letters count as the same character. Store the results in an array with 26 upper- or lowercase character keys and display the contents of the array when all characters have been read from the file. If you like, you can make this an application that runs from a command line, so you can specify the name of the input text file when you execute the application.

12. Write an HTML/PHP application that will calculate and display a monthly loan repayment schedule. The user specifies the loan amount, the annual interest rate, and the duration of the loan in years. Payments are made monthly.

      For n loan payments, where n is the number of years times 12, the monthly payment P for a loan amount A at annual interest rate r (expressed as a decimal fraction, not a percent) is

$$P = (A \cdot r/12)/[1 - 1/(1 + r/12)^n]$$

At the end of the loan repayment schedule, display the total amount received in loan payments.

Suppose you were thinking about lending this money yourself. The alternative is to deposit the money in an interest-bearing account. What APY (annual percent yield) would that account have to pay in order for you to have the same amount of money at the end of y years as you would have received from the loan repayments?

If you don't reinvest the loan payments as you receive them, calculate the APY from:

$$A_{final} = A_{start} \cdot (1 + r_{APY})^y$$

If you immediately reinvest each loan payment in an account paying an annual rate R (presumably lower than rate r) then at the end of y years (n months) that account will hold

$$A_{final} = A_{start} \cdot [(1 + R/12)^n - 1]/(R/12)$$

Here is an example. The monthly payments for a two-year, 8% loan of $200,000 are $9045.46. The total amount paid is $24 \times \$9045.46 = \$217,091$. The APY for an account with an initial deposit of $200,000 that would yield this amount is $(A_{final}/A_{start})^{(1/y)} - 1 = 4.19\%$. Suppose you reinvest the monthly payments as you receive them at 4%, compounded monthly. When the loan is repaid, you will have a total of $225,620, which is equivalent to an APY of 6.21% on a two-year investment of the $200,000.

		Payment	Balance	Reinvestment
			200000.00	Rate = 4%
Payment #	1	9045.46	192287.88	9045.46
	2	9045.46	184524.34	18121.07
	3	9045.46	176709.04	27226.93
	4	9045.46	168841.64	36363.14
	5	9045.46	160921.79	45529.81
	6	9045.46	152949.15	54727.04
	7	9045.46	144923.35	63954.92
	8	9045.46	136844.05	73213.56
	9	9045.46	128710.88	82503.06

10	9045.46	120523.50	91823.53
11	9045.46	112281.53	101175.07
12	9045.46	103984.61	110557.78
13	9045.46	95632.39	119971.76
14	9045.46	87224.48	129417.13
15	9045.46	78760.52	138893.98
16	9045.46	70240.13	148402.41
17	9045.46	61662.94	157942.55
18	9045.46	53028.57	167514.48
19	9045.46	44336.63	177118.32
20	9045.46	35586.75	186754.17
21	9045.46	26778.54	196422.14
22	9045.46	17911.60	206122.34
23	9045.46	8985.55	215854.88
24	9045.46	0.00	225619.85
Total Income	217091.00		225619.85
Return	4.19%		6.21%

13. Modify Document 3.13 (`cardShuffle.php`) so that the code will display four shuffled "hands" of 13 cards each, identified by value and suit:

Three of Clubs
King of Clubs
...
Ten of Spades
Deuce of Spades

14. Define a "heat wave" as a condition for which the maximum temperature exceeds 90°F on any three consecutive days. Write a PHP application that will read and display a file of daily maximum high temperatures, including in your output an appropriate message when a heat wave is in progress.

   Note that you can define a heat wave only retroactively, because the heat wave is known to be occurring only on the third day. This means that you must store data from at least the two previous days before you can display an appropriate message for the heat wave days.

   Here is a sample data file with appropriate output:

07/01/2006 89
07/02/2006 90 heat wave day 1

07/03/2006 93 heat wave day 2
07/04/2006 92 heat wave day 3
07/05/2006 94 heat wave day 4
07/06/2006 89
07/08/2006 91 heat wave day 1
07/09/2006 90 heat wave day 2
07/10/2006 92 heat wave day 3
07/11/2006 89
07/12/2006 87

15. The value of equipment used in manufacturing and other businesses declines as the equipment ages. Businesses must recover the cost of "durable" equipment by depreciating its value over an assumed useful lifetime of n years. At the end of n years, the equipment may have either no value or some small salvage value. Depreciation can be computed three ways:

1. *Straight-line depreciation.* The value of an asset minus its salvage value depreciates by the same amount each year over its useful life of n years.

2. *Double-declining depreciation.* Each year, the original value of an asset minus the previously declared depreciation is diminished by 2/n. (This method does not depend on an assumed salvage value.)

3. *Sum-of-digits depreciation.* Add the integers from 1 through n. For year i, the depreciation allowed is the original value of the asset minus its salvage value, times $(n - i) + 1$, divided by the sum of the digits.

Write an HTML document that allows the user to enter the original value of an asset, the number of years over which the depreciation will be taken, and its salvage value at the end of the depreciation period. Then write a PHP application that will use these values to print out a depreciation table showing the results for each depreciation method. Here is a sample table.

The code that generated this table used echo statements and the round() function to generate the output, because that was a little easier to do while the code was being developed. You can gain more control over the output by, for example, having 100 print as 100.00, using printf() with appropriate format specifiers.

Businesses often like to "front load" the depreciation of an asset in order to realize the maximum tax deduction in the year that the funds were actually spent for the equipment. For this reason, they would likely not choose the straight line method even though it is the simplest of the three.

Original value	$1000					
Salvage value	$100					
Lifetime (years)	7					
Year	Straight line	Asset value	Double declining	Asset value	Sum of digits	Asset value
1	128.57	871.43	285.71	714.29	225	775
2	128.57	742.86	204.08	510.2	192.86	582.14
3	128.57	614.29	145.77	364.43	160.71	421.43
4	128.57	485.71	104.12	260.31	128.57	292.86
5	128.57	357.14	74.37	185.93	96.43	196.43
6	128.57	228.57	53.12	132.81	64.29	132.14
7	128.57	100	37.95	94.86	32.14	100

16. When analyzing a time sequence of measurements made on a noisy system, it is often useful to smooth the data so that trends are easier to spot. One simple smoothing technique is a so-called unweighted moving average. Suppose a data set consists of n values. These data can be smoothed by taking a moving average of m points, where m is some number significantly less than n. The average is unweighted because old values count just as much as newer values. The formula for calculating the smoothed average value S_i corresponding to the ith value in the data set, is

$$S_i = \frac{\left(\sum_{j=i-m+1}^{i} x_j \right)}{m}, \quad i \geq m$$

Given a set of n points, an algorithm for calculating a moving average of m values is:

1. Calculate the sum s of the first m points. The first average (S_m) equals s/m.

2. For each value of i = m + 1 to n, add the ith value to s and subtract the $(i - m)^{th}$ value. Then calculate the average for this new sum.

3. Repeat step 2 until i = n.

Write a PHP application that reads a file of numerical values and creates a new file containing the original values and the moving average smoothed values. You may wish to create an HTML interface that specifies the file name and the number of points to be included in the moving average.

17. The orbit of a body rotating around a gravitational center is characterized by its orbital period τ, the time required to complete one complete revolution starting at perigee (the closest approach to

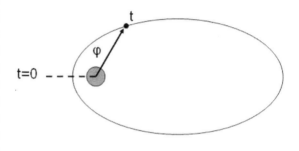

the gravitational center), and its eccentricity e, the departure from a circular orbit. Eccentricity is a dimensionless quantity between 0, for a circular orbit, and 1, when the orbit becomes a parabola. For intermediate values of e, the orbit is an ellipse with the gravitational center at one focus of the ellipse. The speed of an orbiting object is maximum at its perigee (t = 0), and minimum at its apogee (t = τ/2).

For a circular orbit, the angular position φ of an object in its orbit is simply related to time t ≤ τ:

$\varphi_{e=0} = 2\pi(t/\tau)$ radians

When the orbit is elliptical, the calculation of the angular position of an object in its orbit is much more complicated. The "mean anomaly" M for any orbit is the same as the true angular position for a circular orbit with the same period:

$M = 2\pi(t/\tau)$ radians

But, the actual angular position φ, the "true anomaly," for a non-circular orbit cannot be calculated directly. The mean anomaly is related to the so-called eccentric anomaly E_c through a transcendental equation:

$$M = E_c - e \cdot \sin(E_c)$$

After E_c is found, then the true anomaly φ can be calculated directly:

$$\varphi_{0<e<1} = \arccos\{[\cos(E_c) - e]/[1 - e \cdot \cos(E_c)]\}$$

If M is greater than π radians (that is, if t/τ > 0.5), then let φ = 2π − φ. Express your final answer in degrees (degrees = radians·180/π).

To solve for E_c using a recursive function with t, τ, e, and the current guess for E_c as the four parameters:

1. As an initial guess, let E_c = M and use this value in the initial call to the function.

2. In the function, calculate $newE_c = M + e \cdot \sin(E_c)$.

3. Recursively call the function with $newE_c$ as the fourth argument.

4. Keep recalculating until the absolute value $|newE_c - E_c|$ is less than some suitably small value—1×10^{-5} is a reasonable choice.

When the terminating condition for recursive calls is satisfied, then calculate φ as defined. Use an HTML interface something like this:

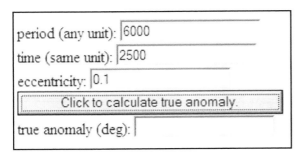

Values of eccentricity very to close to 1 may cause numerical problems, including failure of recursive functions to reach their terminating condition. If eccentricity is equal to 1, then your code should display an appropriate message—that the object's path is a parabola and not an ellipse. Eccentricity values greater than 1 or less than 0 are simply not allowed as input.

18. Rewrite the PHP application in Document 3.14b so that it does not store data from the file in an array for any of the operations specified in the problem statement.

19. Using Documents 5.3a and 5.3b as a guide, write a PHP application that accepts as input the "weather report" barometric pressure and returns the "station" pressure. Except at a few research sites, the reported barometric pressure value is always corrected to sea level—otherwise it would not be possible to understand maps of high and low pressure associated with weather systems. The actual barometric pressure at a site (station pressure) can be obtained from the reported pressure by adjusting it for site elevation h, in units of kilometers:

$$p_{station} = p_{sea\ level} \cdot exp(-0.119h - 0.0013h^2)$$

In the United States, barometric pressure is reported in units of inches of mercury. Almost everywhere else in the world, the units are millibars (hectopascals). Values for standard atmospheric conditions at sea level are 1013.25 millibars or 29.921 inches of mercury. Because these two units have such different values associated with them, your code can determine the units in which the pressure was entered. If the sea level pressure value entered is less than 40, assume the units are inches of mercury and convert that value to millibars:

$$p_{millibars} = p_{inches\ of\ mercury} \cdot (1013.25/29.921)$$

As an example, look at an online weather report for Denver, Colorado, USA. Denver is often called the "mile high city" because its elevation is about 5300 ft (1.6 km). The barometric pressure will be reported as a value typically just a little above 1000 millibars, just as it is at sea level. Use your PHP application to calculate the actual barometric pressure in Denver under standard atmospheric conditions.

Glossary

The Chapter and section in which a term first appears is given in parentheses following the term. The first appearance in the text is printed in bold font.

append (text file) (1.1)
A "write-only" access permission that allows new information to be appended to the end of an existing text file.

ASCII character sequence (3.2)
A standardized representation of characters.

client-side language (1.1)
A programming language such as JavaScript that resides on a local browser and can process scripts downloaded to the browser, as opposed to a server-side language such as PHP.

command line interface (CLI) (5.1)
A text-based computer interface that allows a user to type commands, enter data from the keyboard, and display text output from a program.

constructor (3.1)
A means of defining the properties and contents of a built-in or user-defined data object, such as the array constructor in PHP.

escape character (4.4)
A backslash (\), indicating that the following character has a special meaning.

escape sequence (4.4)
A backslash followed by a character.

file handle (1.1)
The "logical" name by which a physical file is identified within a program.

file name extension (1.1)

A set of (usually) three or four characters following a period (.) which identifies the nature of a file and its contents. File extensions are often associated with specific applications, such as .php for any file containing text that can be interpreted as a PHP script.

floating-point number (1.1)

A real number, and a particular way of representing such a number within computer memory. Whole numbers can be represented as either integers or floating-point numbers.

format (1.1)

A specification for reading or writing data from or to a file or other resource, such as a keyboard, or a description of how the contents of a file are organized.

format string (1.1)

A string that provides information about the contents and format of values in a file.

header line(s) (1.1)

One or more lines in a file which identify and/or describe the contents of that file, as opposed to the data themselves.

language construct (1.1)

A reserved term or group of terms in a programming language which performs certain operations.

local computer (server) (1.1)

In this book, the term "local computer" usually refers to a user's own computer that is also running a server application.

PHP document (1.1)

Any text document that can be interpreted as a PHP script.

PHP environment (1.1)

A local or remote server that includes a PHP script interpreter, a place to store PHP scripts, and a place where files can be created, read, and modified.

PHP interpreter (1.1)
>A computer application that interprets PHP script files.

PHP script (1.1)
>A series of statements that follow PHP syntax rules, and that can be executed by a PHP interpreter.

PHP tag (1.1)
><$php... $>, an HTML tag which contains PHP statements.

pseudo data type (4.1)
>A type specifier such as (mixed) used to indicate the data type or types associated with a variable name or other identifier.

random access (4.4)
>Pertaining to a data file whose contents can be accessed in any order.

read-only (text file) (1.1)
>A text file available for access from within a PHP script in a way that only allows its contents to be read but not modified in any way.

remote server (1.1)
>A server running somewhere other than on a local computer.

resource (4.1)
>Any data source, such as a data file or a keyboard, that is external to but accessible to a PHP application. Resources are represented by the pseudo data type (resource).

sequential access (4.4)
>Pertaining to a data file whose contents can be accessed only sequentially, starting at the beginning.

server (1.1)
>A software application that provides services such as file access to other computer programs and users on the same computer (a local server) or some other computer (a remote server). A computer on which a server application is running is often referred to as a server even if it is also used for other purposes.

server side (1.1)

Refers to activities taking place on a server or data files residing in folders accessible through a server, even if that server is on a user's local computer.

server-side language (1.1)

A programming language such as PHP which resides on a computer server, as opposed to being available within a local (client-side) browser.

write-only (text file) (1.1)

A text file that be created or overwritten from within a PHP script.

Index

NOTE: For HTML tags and PHP functions and constructs, only first appearances or significant new uses are indexed. PHP math functions are not indexed individually; tables of all math functions are found at the index reference to "math functions" and "functions, math".

$ (symbol in variable name) 38
$_POST[...] 5, 11, 37, 60
=> (key association operator) 49
. operator 49

A
"a" (file access mode) 17
AceHTML code editor vii, 7
action="..." 3
alert(...) 44
alignment specifier 84
An Introduction to HTML and JavaScript for Scientists and Engineers v
append permission 17
application
  PHP 4
  server-side 3
argument passing 19
array 49
  constructor 39, 49
  constructs 99
  data type 76
  element 49
  functions 99
  index 49
  key 49
  sorting 54
  two-dimensional
array(...) 99
array_keys(...) 60, 99
array_pop(...) 57, 99
array_push(...) 57, 100
array_shift(...) 57, 101

array_unshift(...) 57, 101
arrays, two-dimensional 52
ASCII character sequence 54, 117

B
batch processing 24
<body> 3
boolean data type 75
break 78, 102
browser 8

C
calculations, PHP 20
card shuffle problem 68
case 78
case-controlled conditional execution 78
character sequence, ASCII 54, 117
characters, escape 86
checkbox values, reading 62
chr(...) 94
client-side language 1
cloud observation problem 62
code editor, AceHTML vii, 7
command line 24
  environment and interface107
comments, PHP 23
computer, local 1
conditional execution 77
conditional execution, case-controlled 78
conditional loop 47
condition-controlled loops 82
construct, language 9

constructor, array   39, 49
constructs   83
  array   99
  miscellaneous   102
copy (...)   102
count (...)   53, 102
count-controlled loops   80

D
data file management   69
data structure, user-defined   49
data type   75
  string   75
  array   76
  boolean   75
  float   75
  integer   75
  mixed   76
  number   76
  pseudo   76
  resource   76
declaration, variable   23
die (...)   103
do...   82
document, HTML   3, 19
Documentation Group, PHP   v
documents, PHP   7
dot operator (.)   49

E
echo...   5, 8
editor, AceHTML   vii, 7
element, array   49
else...   5, 14
elseif...   14
environment, command line   107
environment, PHP   6
escape characters   86
eval (...)   48
execution, conditional   77
exit ()   11, 103
extension, file name   16

F
fclose (...)   5, 12, 87

feof (...)   5, 12, 42, 87
fgetc (...)   88
fgets (...)   5, 12, 42, 88
FIFO   57
file
  access modes   87
  handle   17, 86
  handling   84
  I/O, PHP   40
  name extension   16
  sharing   71
  logical name   86
  physical name   86
  server-side   15
file (...)   88
first-in-first-out data storage   57
float data type   75
floating point number   13
fopen (...)   5, 11, 86
for... loop   80
foreach... loop   81
<form>   3
format
  specification   13
  specifier   17, 84
  format string   13
fprint (...)   16, 43, 89
fread (...)   90
fscanf (...)   12, 42, 90
function ...   38
function, user-defined   38
functions
  array   99
  I/O   84
  math   96
  miscellaneous   102
  PHP   19, 23, 83

G
graphical user interface   112
GUI   112

H
<h2>   3
handle, file   17, 86

`<head>` 3
histogram problem 66
`<html>` 3
HTML document 3, 19
HTML/JavaScript 1

I
I/O functions 84
`if (...)` 5
if... then... else... 13, 77
IIS server 7
increment/decrement operators 76
index, array 49
information, passing 10
`<input>` 3, 10
input, keyboard 111
integer data type 75
interface
  command line 107
  graphical user 112
interpreter 6

J
JavaScript 1
`.js` 1

K
key
  array 49
  starting 52
keyboard input 111
keys
  named 50
  numerical 49
  sequential 50
keyword 23

L
language construct 9
language
  client-side 1
  compiled 6
  interpreted 6
  server-side 2
last-in-first-out data storage 57

Lerdorf, Rasmus v
LIFO 57
line feed characters 43
`list (...)` 5, 12, 39, 105
literal, string 9
local computer 1
`localhost` 7
logical operators 76
loop 79
  conditional 47
  condition-controlled 82
  count-controlled 80
  `for...` 80
  `foreach...` 81

M
management, data file 69
math functions 96
`method="get"` 10
`method="post"` 3
miscellaneous
  constructs 102
  functions 102
mixed data type 76
modes, file access 87
multiple form submissions,
preventing 20

N
`name="..."` 3
named keys 50
number data type 76
number, floating point 13
numerical keys 49

O
object mass problem 43
`onclick` 44
operator precedence 76
operator, . (dot) 49
operators 75
  increment/decrement 76
  logical 76
  relational 76
`<option>` 44

ord(...) 94

P

<p> 3
padding specifier 84
parseFloat() 24
parseInt() 24
passing information 10
passing, argument 19
permission, append 17
  read 12
  write 17
Personal Home Page Tools v
<?php...?> 5
PHP v, 1
  applications 4
  calculations 20
  comments 23
  Documentation Group v
  documents 7
  environment 6
  file I/O 40
  functions 19, 23
  interpreter 6
  script 9, 19, 23
  tag 9
PHP and JavaScript 36
  differences 36
phpinfo(...) 9
precedence, operator 76
precision specifier 85
preventing multiple form
submissions 20
printf(...) 90
printf_r(...) 92
probability density function 108
pseudo data type 76

Q

quadratic equation problem 18, 59
queues 56

R

"r" (file access mode) 12
read permission 12, 17

reading checkbox values 62
read-only text file 12
relational operators 76
remote server 1
rename(...) 102
resource data type 76
restrictions, server-side language
24
return 39
round() 38

S

<script> 3
script, PHP 9, 19, 23
<select> 44
sequential keys 50
server 1
server
  IIT 7
  remote 1
  Windows IIS 7
server-side
  application 3
  file 15
  language 2
  language restrictions 24
sharing, file 71
sign specifier 84
similarities, PHP and JavaScript
36
sizeof(...) 53
sizeof(...) 102
sort(...) 54
sorting, array 54
space-separated text file 3
specification, format 13, 17, 84
sprintf(...) 93
sscanf(...) 42, 93
stacks 56
starting key 52
STDIN 111
strcasecmp(...) 94
strcmp(...) 94
string data type 75
string

literal   9
format   13
strlen (...)   5, 11, 95
strncasecmp (...)   12, 95
strncasecmp (...)   95
strncmp (...)   95
strtolower (...)   96
strtotime (...)   70, 104
strtoupper (...)   96
switch (...)   78

T
tag, PHP   9
test file
  space-separated   3
  read-only   12
<title>   3
two-dimensional arrays   52
type specifier   85
type="button"   44
type="radio"   62
type="submit"   10
type="submit"   32
type="text"   3
types, data   75

U
URL   7
user-defined
  data structure   49
  function   38

V
.value (JavaScript)   24
value="..."   4
var (JavaScript)   23
var_dump (...)   105
variable declaration   23
variable names   38
Visicomm Media   vii
vprintf (...)   93

W, X, Y, Z
"w" (file access mode)   17
water vapor problem   2, 24
while {...}   5, 12, 42, 82
width specifier   84
Windows IIS server   7
write permission   17
wwwroot   7

Printed in the United States